Designing Solid-State Power Supplies

Designing Solid-State Power Supplies

Robert G. Seippel
Instructor, Electronics
College of the Canyons, Valencia, California

Roger Lincoln Nelson
Development Engineer
Honeywell Corporation
Minneapolis, Minnesota

American Technical Society ■ Chicago 60637

Library of Congress Catalog Number: 74-31519
ISBN 0-8269-1591-4

123456789 75 987654321

Printed in the United States of America

PREFACE

In his efforts to become an electronic technician or engineer, the student usually passes through the power supply sequence. In all electronic programs the knowledge of power supplies is a requirement.

The subject of power supplies is usually studied between ac/dc fundamentals and solid-state devices. Since power supplies are absolutely essential in electronic circuits, a study of them as a separate entity is appropriate.

The purpose of this book is to provide the design requirements for power supplies in a different and simplified manner. The early part of the book covers the current flow analysis, preparing the reader for a brief survey of practical power-pack designs. Following this, the text explains how to choose components. Actual design of power supplies then follows in cookbook style, providing the reader with a step-by-step procedure and a design example. Final sections of the text cover trouble-shooting and procedures for testing power supplies.

An appendix is included covering the subject of electrical/electronic safety, another illustrates typical rectifier devices and their specifications, and a final appendix shows some of the available solid-state power supplies.

The writer believes that this book, by means of its unusual and simple method of presentation, will provide the needed understanding of how to build a power supply. The text can therefore be used by students of electronics as a learning device and by experienced technicians as a quick refresher.

Robert G. Seippel

THANKS TO:

Hazel, my wife—Typist
Dal Fitts—Technical Editor

TABLE OF CONTENTS

List of Illustrations

List of Tables

Section 1

Circuit Analysis of Solid State Rectifier Circuits

Solid state power supplies are replacing those using tubes. The reasons for this are obvious: there is no warm-up period or filament requirement for heating, power losses are low, and the newer power packs are smaller. Also, since the state of the art has expanded, prices for them are lower.

Semiconductor power supplies use the silicon rectifier, which is light, reasonably priced, and can be stacked in series to handle higher ac inputs. In fact, the solid state rectifier can replace the electron tube diode in almost any modern application.

The circuitry for constructing solid state power supplies is simple. Since no filament voltage is required, one of the basic power supply trouble areas is eliminated. While it is true that heat can be a problem with solid state power rectifiers, with proper heat sinks this problem is reduced to a minimum. Transients during input or output switching or load changes may also cause solid state unreliability, but the transient problem may be eliminated by proper circuit design.

HALF-WAVE RECTIFIER CIRCUIT (SEE FIG. 1.)

The half-wave rectifier is the simplest of all the power supplies. It consists of a transformer and one rectifier diode. The transformer is a step-down type because power for the circuit usually comes from a 115 vac or a 220 vac source. Solid state circuits are typically biased by low voltage dc (example 15 vdc). Therefore the ac voltage must be transformed to a low level prior to rectification.

Transformers for power supplies are chosen to have a secondary average voltage output that is near the dc level required after rectification. Transformers are also chosen to be able to withstand the current demands of the load.

Diodes are chosen for their current capability and peak inverse voltage rating (PIV). Peak inverse voltage is defined as the maximum reverse-bias voltage that may be accepted by the rectifier. This voltage may vary somewhat with temperature. Actual construction of the power supply requires that the diode be installed

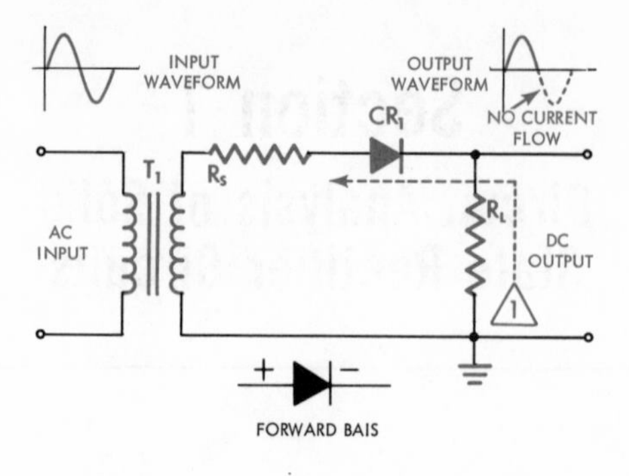

ANALYSIS:

1. POSITIVE ALTERNATION of input sine wave forward biases CR1 and electron current will flow through the diode and load.
2. NEGATIVE ALTERNATION of input sine wave reverse biases CR1 and electron current stops flowing.

Fig. 1. Half-Wave Rectifier Circuit

on a heat sink to help dissipate some of the heat generated by current flow.

The half-wave rectifier is not very efficient. The reason for this, is that the average dc derived by the circuit is low. In fact the average voltage is only half that of the full-wave rectifier. Without filtering, the circuit is very inefficient to use under any application. Pulsating dc output ripple is determined by input frequency.

The circuit shown in Fig. 1 is a typical half-wave rectifier. The transformer secondary has one lead tied to reference (ground). The second lead has a resistor R_S in series with the diode. The resistor is a current-limiting resistance. Its purpose is to limit the current flowing through the diode. If the secondary dc resistance in the transformer is great enough, the resistor is not needed.

On the positive alternation of input voltage the diode is forward biased and current flows through the diode and the load. On the negative alternation of the input voltage, the diode is reverse-biased and current stops flowing except for minor reverse current. This arrangement gives pulsating dc output in the positive direction. To obtain a negative dc pulse, the diode is placed in the circuit in the opposite direction.

FULL-WAVE RECTIFIER CIRCUIT (SEE FIG. 2)

The full-wave rectifier is similar in operation to the half-wave rectifier. The transformer utilized has a center tap. Each leg of the

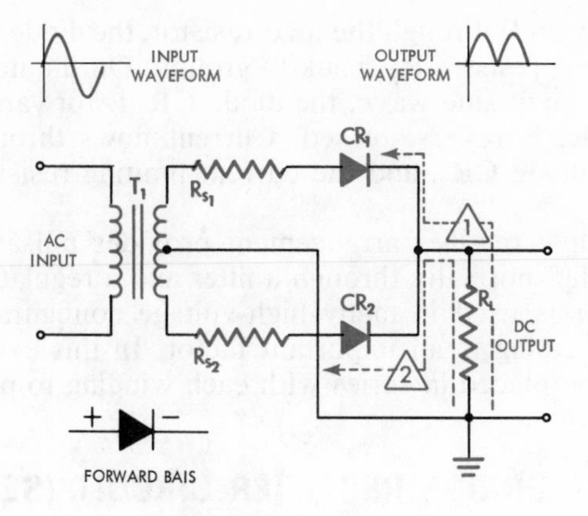

Fig. 2. Full-Wave Rectifier Circuit

center tap should be capable of an average ac voltage near the level expected of the dc level output. The center tap is tied to reference (ground).

Two diodes are used. They are placed in the same direction in series with the transformer secondary winding and series current-limiting resistors. These limiting resistors R_{S1} and R_{S2} are used to limit current through diodes CR_1 and CR_2 if the secondary winding does not have enough resistance.

The full-wave rectifier is much more efficient than the single wave because it utilizes the secondary of the transformer for 360 degrees of the input sine wave. Current flows through one diode for one half cycle and the opposite diode for the second half cycle. Transformer and diodes are chosen in the same manner as in half-wave rectifiers. Diodes in full-wave rectifiers must be mounted on heat sinks as in the single wave configuration.

Output of the full-wave rectifier is usually filtered, then regulated, Pulsating dc output ripple is twice the input frequency.

The circuit shown in Fig. 2 is a typical full-wave rectifier. On the positive alternation of the input sine wave, the diode CR_1 is forward-biased and the diode CR_2 is reverse-biased. Current flows from the center tap (ground) through the load resistor, the diode CR_1, and the current limiting resistor R_{S1} back to ground. On the negative alternation of the input sine wave, the diode CR_2 is forward-biased and the diode CR_1 is reverse-biased. Current flows through the load resistor, the diode CR_2, and the current limiting resistor R_{S2} back to ground.

The full-wave rectifier arrangement provides pulsating dc. The dc power is fed normally through a filter and a regulator. The full-wave rectifier is used in many high-voltage configurations where peak inverse rating is an important factor. In this event a pair of diodes may be placed in series with each winding to prevent rectifier breakdown.

FULL-WAVE BRIDGE RECTIFIER CIRCUIT (SEE FIG. 3.)

The bridge rectifier is a full-wave rectifier. It relies on two rectifiers, one on each arm of the transformer, to conduct together on each alternation. The full-wave bridge rectifier does not require a center tap as does the conventional full-wave rectifier. It conducts during the full secondary output of the transformer. There are four diodes used in the bridge. Each diode is connected in series with and operates at the same time as the diode on the opposite transformer winding.

In Fig. 3 the reader will see that CR_1 and CR_3 are forward-biased when CR_2 and CR_4 are reverse-biased. This arrangement allows the bridge rectifier to require only one half the peak inverse rating of the conventional full-wave rectifier during conduction. The other advantage is that the bridge rectifier utilizes the entire secondary voltage at all times, making it more efficient. The current-limiting resistor R_S is used to limit current through the diodes in the event of a short. If the transformer secondary has enough resistance, this resistor is not necessary.

The circuit shown in Fig. 3 is a typical full-wave bridge rectifier. On the positive alternation, the diodes CR_1 and CR_3 are forward-biased. The diodes CR_2 and CR_4 are reversed biased. Current flows from ground through diode CR_1, through the current limiting resistor R_S, the transformer secondary, the diode CR_3, and the load, back to ground.

On the negative alternation, the diodes CR_2 and CR_4 are forward-biased. The diodes CR_1 and CR_3 are reverse-biased. Current flows from ground through the diode CR_4, the transformer secondary, the current limiting-resistor R_S, the diode CR_2, and the load, back to ground. This arrangement provides pulsating dc voltage.

The dc is normally fed through a filter and a regulator. The bridge rectifier can be designed with a center-tap transformer to provide two load voltages simultaneously. This circuit is seldom used, however, because of the current drain.

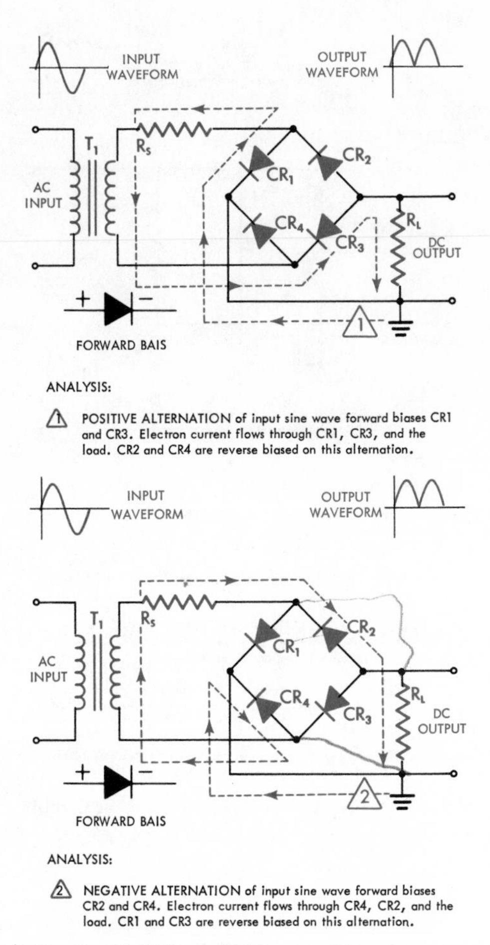

Fig. 3. Full-Wave Bridge Rectifier Circuit

FULL-WAVE VOLTAGE DOUBLER CIRCUIT (SEE FIG. 4.)

The full wave voltage doubler consists of a full wave rectifier with parallel capacitors and bleeder resistors. The capacitors charge as conduction takes place through the diode at a rate determined by the resistor R_S and the capacitance time constant.

Bleeder resistances R_{B_1} and R_{B_2} determine the discharge time

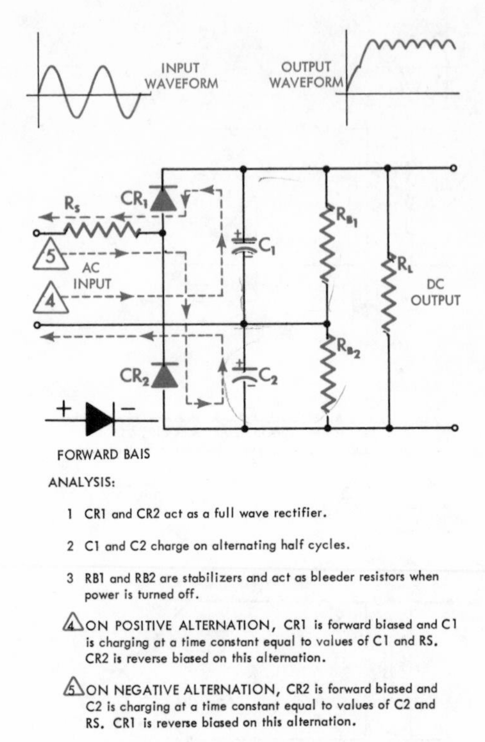

ANALYSIS:

1 CR1 and CR2 act as a full wave rectifier.

2 C1 and C2 charge on alternating half cycles.

3 RB1 and RB2 are stabilizers and act as bleeder resistors when power is turned off.

4 ON POSITIVE ALTERNATION, CR1 is forward biased and C1 is charging at a time constant equal to values of C1 and RS. CR2 is reverse biased on this alternation.

5 ON NEGATIVE ALTERNATION, CR2 is forward biased and C2 is charging at a time constant equal to values of C2 and RS. CR1 is reverse biased on this alternation.

6 RB1 and RB2 are very large values (2 or 3 megohm) therefore, C1 and C2 discharge very little during alternate cycles.

Fig. 4. Full-Wave Voltage Doubler Circuit

constant. They are normally very large (two or three megohms). All components are of equal value for equalization. The reason for this is that the capacitor must not discharge too much during the alternate cycle. It must keep its charge because the output voltage is felt in parallel to both capacitors.

Voltage triplers and voltage quadruplers are also available. These two circuits are used in small current applications and are very inefficient because of power losses and very poor regulation.

The circuit in Fig. 4 requires the following analysis. On the positive alternation of the input sine wave, diode CR_1 is forward-biased and diode CR_2 is reverse-biased. Current flows through the diode CR_1 and the capacitor C_1 is charging at a time constant $TC = RC$ (where TC = time constant; R = resistance; and C = capacitance) equal to the values of C_1 and resistor R_S.

On the negative alternation of the input sine wave diode CR_2 is forward biased and diode CR_1 is reverse-biased. Current flows through the diode CR_2 and the capacitor C_2 is charging at a time constant (TC = RC) equal to the values of C_2 and the Resistor R_S. During the negative alternation the capacitor C_1 tries to discharge but can't because of the large value of R_{B1}. During the positive alternation the capacitor C_2 tries to discharge but can't because of the large value of R_{B2}. The load is in parallel with the bleeder resistances, therefore the doubling action is felt as load voltage. If a meter were placed across the load, the meter would be indicating the charge on the two capacitors.

THREE-PHASE HALF-WAVE WYE RECTIFIER CIRCUIT

(SEE FIG. 5.)

The three-phase wye rectifier illustrated in Fig. 5 is the basic three-phase rectifier circuit. This circuit consists of a single-wave rectifier for each secondary leg (e_{sec}).

Each leg has voltage induced from the delta primary that is 120 degrees phase-shifted from the other two secondary legs. Thus each rectifier conducts for 120 degrees, or one third of input cycle. Electron current flow is in pulses for every other half cycle, and current flows through the load as each leg rectifies. Ripple voltage is three times the frequency of the source ac voltage, therefore less filtering is required.

The primary windings of the transformer T_1 are delta-wound; however they could be wye-wound, depending on design requirements. The load is taken at a junction of the three phases (common point) and a common junction that ties the three rectifiers together.

The output is generally configured in this manner. If all the diodes were reversed, the rectifier circuit would become a negative dc

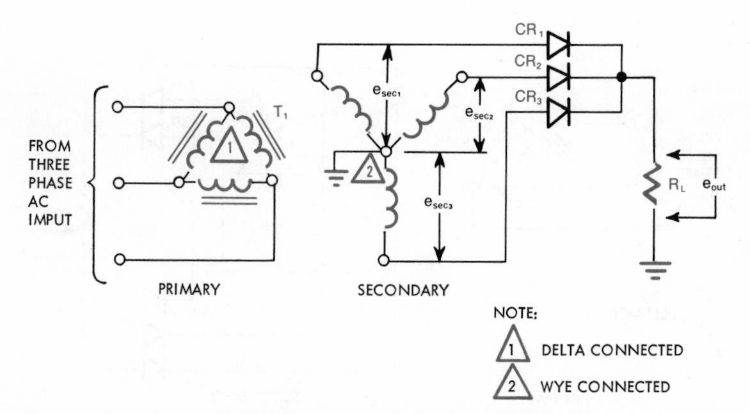

Fig. 5. Three Phase, Half-Wave Wye Rectifier Circuit

power supply. In some cases the rectifiers as shown are representative of a series of two or three rectifiers, since the reverse voltage rating in any secondary leg must exceed the maximum peak-inverse voltage seen in the entire circuit.

THREE-PHASE FULL-WAVE WYE SECONDARY RECTIFIER CIRCUIT (SEE FIG. 6.)

The three phase wye rectifier illustrated in Fig. 6 is similar to the single half-wave rectifier previously discussed. The circuit consists of a pair of single-wave rectifiers for each secondary leg. One rectifier is used for positive alternation rectification and other (placed in opposite direction) is used for negative alternation rectification.

Each of the six rectifiers conduct for 120 degrees of the input cycle. The sequence of conduction has a positive and a negative rectifier conducting at the same time, hence an overlap of conducting period.

In Fig. 6, the conduction sequence is CR_1 and CR_4, CR_1 and CR_5, CR_2 and CR_5, CR_2 and CR_6, CR_3 and CR_6, and CR_3 and CR_4, then repeated in the same sequence. At any given time current flows through the load. The frequency ripple of the output is six times the frequency of the ac source voltage, therefore very little filtering is required.

The primary windings of the transformer T_1 are delta-wound, however they could be wye-wound, depending on design requirements. The output voltage is taken from a junction of the three phases (common point) and a common junction that ties the six rectifier outputs together.

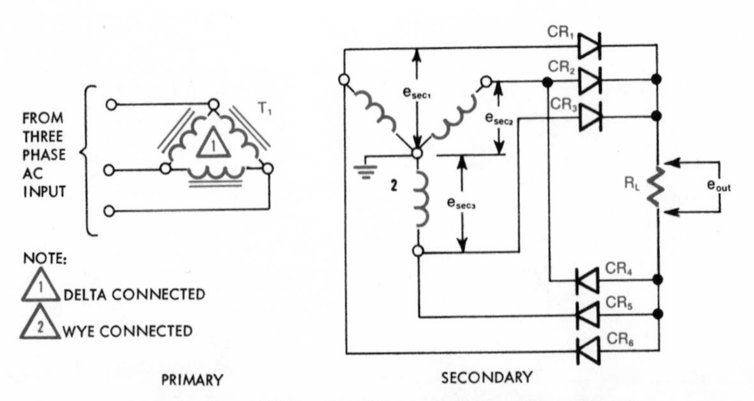

Fig. 6. Three Phase Full-Wave Wye Rectifier Circuit

The output of this rectifier is generally configured in this manner. In some cases, the rectifiers as shown are representative of a series of two or three rectifiers. The peak inverse voltage across a single rectifier is 2.54 times the rms volts across one secondary phase winding.

THREE-PHASE FULL-WAVE DELTA SECONDARY RECTIFIER CIRCUIT (SEE FIG. 7.)

The full-wave delta power supply is used whenever massive amounts of power are required by the load. The primary and the secondary of the power transformer are both delta-connected. The three-phase transformer is stepped up to the secondary voltage required by the load. At any one time current flows through the load.

Each of the diodes conduct for 120 degrees of the input electrical cycle, as each secondary winding is 120 degrees out of phase with the other two windings. A pair of diodes are connected opposite to one another from each terminal of the secondary so that current will flow from that terminal during the positive and negative cycles of the sine wave. There is an overlap of rectifier conduction.

The rectifiers conduct in pairs as determined by phasing of the transformer secondary voltages. The rectifier conduction pairs are thus: CR_1 and CR_6, CR_2 and CR_6, CR_2 and CR_4, CR_3 and CR_4, CR_3 and CR_5, CR_1 and CR_5, then CR_1 and CR_6 again.

The ripple frequency of the output voltage is six times the ac source frequency, therefore very little filtering is required. The peak inverse rating of each individual rectifier is approximately 1.42 times the rms voltage across a single secondary winding. Regulation is good to excellent.

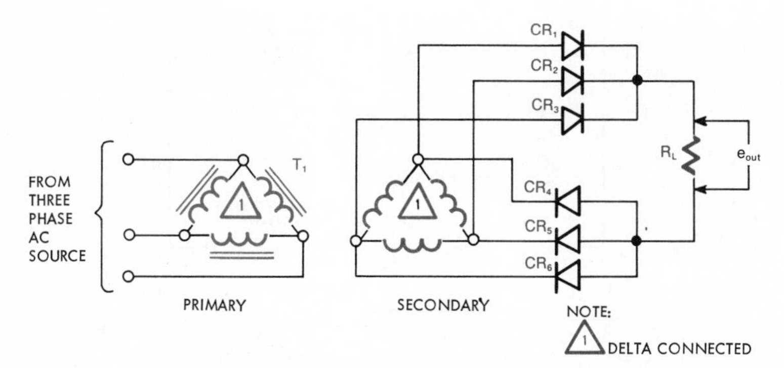

Fig. 7. Three Phase, Full-Wave Delta Secondary Rectifier Circuit

Section 2

Filter Circuits

Filtering is a *must* for the solid state dc power supply. The filter comes in many forms and arrangements and is used to smooth out the ripple voltage caused by rectification. Use of the filter is dependant on space, dc ripple requirements, and cost.

If a capacitor is used for filtering it is placed in parallel with the

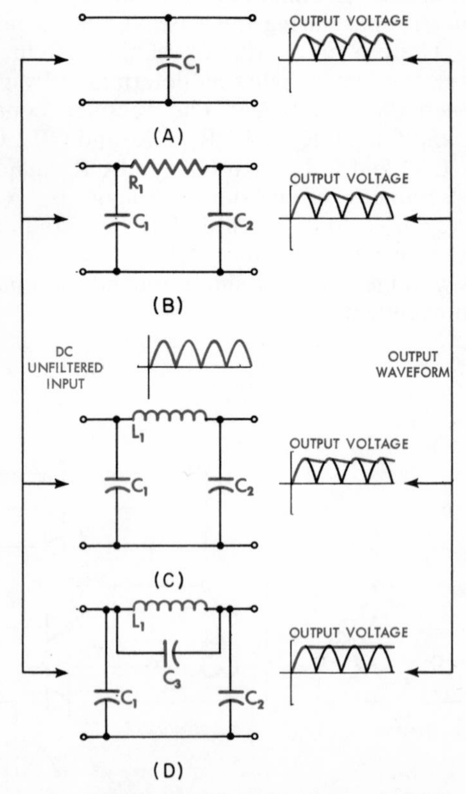

Fig. 8. Typical Filters for DC Power Supplies

load. As you know, the capacitor opposes a change in voltage, so a constant voltage is seen on the load. The larger the capacitor is, the better the filtering capability. However, this causes the capacitor to be heavy and costly.

The choke (inductor) is used in series with the load. As you know, the choke (inductor) opposes a change in current. The capacitor is sometimes used as a filter by itself. (Figs. 8(A) and 8(B).)

Other arrangements are the *L*-filter using one choke in series and one capacitor in parallel; the *T*-filter using two chokes in series and a capacitor between them in parallel; and the *pi* filter using two capacitors in parallel with a choke in series between them.

Each does a respectable job of opposing current change or voltage change. The choke filter has a disadvantage in that output voltage tends to be lower than with capacitor filters. The capacitor filter draws current. The two together then (Fig. 8(C)) are the most efficient. The best, filter is the tuned filter (Fig. 8(D)). The capacitor and choke (inductor) have equal impedances and produce practically ripple-free dc.

Ripple-free dc is demanded in many solid state circuits. Ripples in power may cause switching transistors or SCR's in computers to fire accidently. Ripple dc can cause noise in audio and radio circuits. In some circuits where exact dc levels are required, ripple voltage may even cause some solid state devices to be destroyed.

DC POWER SUPPLY OUTPUT RIPPLE VOLTAGE (SEE FIG. 9.)

Ripple voltage in a dc power supply is that peak area of the ac input to the power supply that was not removed by rectification or filtering. This voltage is usually very small (less than 10% of the dc) but in many cases it is necessary to remove it or at least be knowledgeable of how much ripple is present in the supply.

Ripple voltage may be measured with an ac voltmeter tied in parallel to the power supply output. Ripple voltage is better mea-

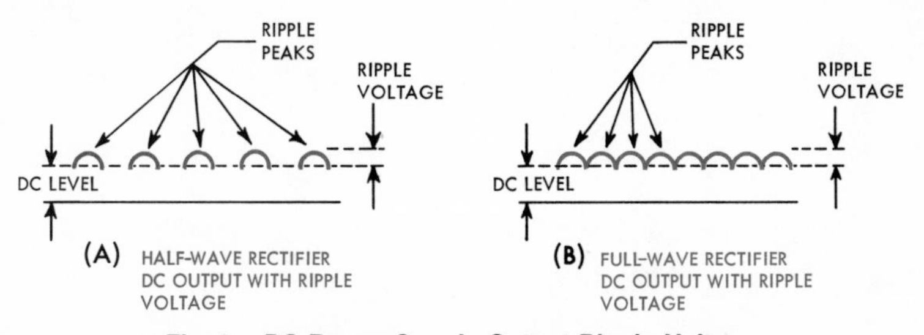

Fig. 9. DC Power Supply Output Ripple Voltage

sured with the use of an oscilloscope. With this instrument an actual picture of the ripple is seen, and fluctuations or distortions in the ripple pattern can be analyzed. A low capacitance probe should be used in measurements.

In Fig. 9, the dc output of a single-wave and a full-wave power supply are shown. The reader will note that the full-wave power supply has two times as many ripples as the half-wave power supply. If the input ac frequency for a half-wave rectifier was 60 hertz, the ripple frequency would have 60 peaks for each second. If the input ac frequency for a full-wave rectifier was 60 hertz, the ripple frequency would have 120 peaks for each second.

Ripple voltage is not measured in peaks, however, but in percentage. Ripple voltage percentage is calculated as a ratio between ac and dc voltage measured in parallel with the power supply, that is: $\text{Percent Ripple} = \frac{\text{ac volts}}{\text{dc volts}} \times 100.$

Section 3
Surge Resistance

In all power supplies there must be a method for limiting current through the rectifiers to a safe value. This is accomplished by either of two basic methods.

The first method is by placing a surge (current-limiting) resistance in series with the semiconductor diode (rectifier). The surge resistance is selected after taking into account the peak current rating of the rectifier, the load, and other circuit specifics such as filtering or regulation.

The second method for limiting-current is by using a transformer with enough resistance in its secondary or a choke input filter in series with the load and the rectifier.

In Fig. 10(A) a single-wave rectifier power supply is shown with a surge resistance in series with the rectifier. In Fig. 10(B) a pair of

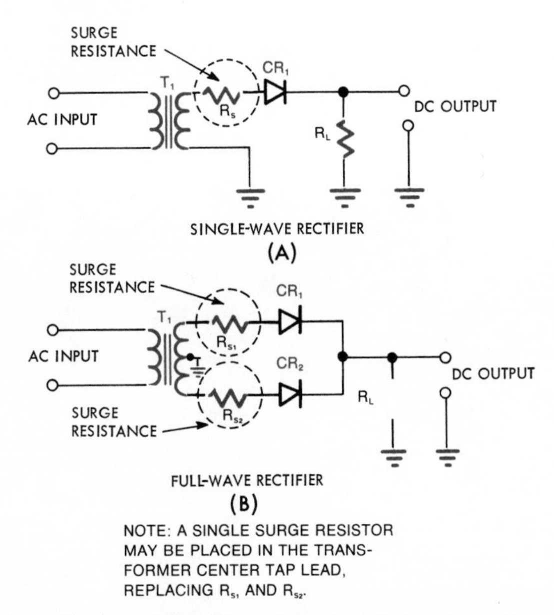

Fig. 10. Single and Full-Wave Rectifiers Using Surge Resistors

surge resistors are used in series with the rectifiers in a full-wave rectifier power supply. A single surge resistance may be used in the ground lead at the transformer center tap in place of the other two resistors. In this manner the surge resistance would be common to both rectifier loops. Surge resistors are most commonly used in high power supplies.

Section 4
Switching Transients

Some of the failures of rectifiers in solid state power supplies are due to switching transients caused by induction in the transformer. When currents in the primary of the power supply transformer are suddenly changed, large transient secondary voltages are developed which result in inverse voltages harmful to the rectifier.

In selenium rectifiers the magnetic energy of the transformer can be dissipated in the rectifier, but in silicon rectifiers the reverse current which causes rectifier breakdown is very low.

In Fig. 11(A) the switch S_1 is opened. The magnetic lines of force that were established in the transformer collapse into the primary windings. This collapse induces a counter voltage into the secondary. Although the time of the change in current is short, the current itself becomes large for that instant, and the transient voltage induced into the secondary has a large enough spike to destroy the silicon rectifier.

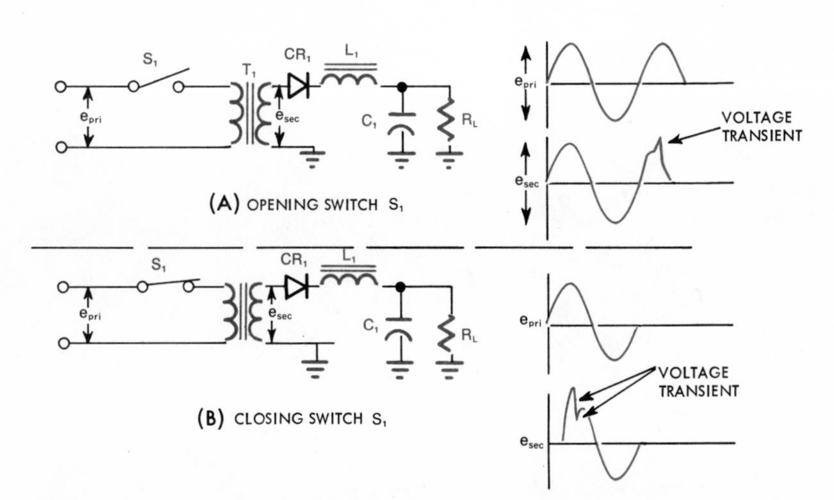

Fig. 11. Voltage Transients in Power Supply Secondary When (A) Opening Primary Switch, and (B) Closing Primary Switch

In Fig. 11(B) the switch is closed. Current immediately flows in the primary of the transformer. This current causes an instantaneous spike transient voltage in the secondary.

After the spike occurs, the current that is set up in the secondary causes a voltage transient in the primary and, in turn, a secondary spike transient. This second spike is smaller than when the switch S_1 is first closed but still greater than the normal secondary peak voltage. The spike can be measured by the use of the relationship $e = L \frac{di}{dt}$, where:

e = Transient voltage

L = Inductance

$\frac{di}{dt}$ = The instantaneous change in current with respect to time.

Section 5
Heat Sinks

Probably the foremost factor contributing to rectifier failure is heat. Any time current flows through a semiconductor, heat is generated. Since rectifiers in power supplies must handle large currents, it is expected that large amounts of heat will be developed. Therefore there must be some method of removing the heat from the semiconductor. It is obvious that heat cannot be eliminated, so the next best thing to do is to dissipate it into the air. This is accomplished by the use of *heat sinks,* which are essentially miniature radiators made of thermally conductive metal. Fig. 12 illustrates typical heat sinks.

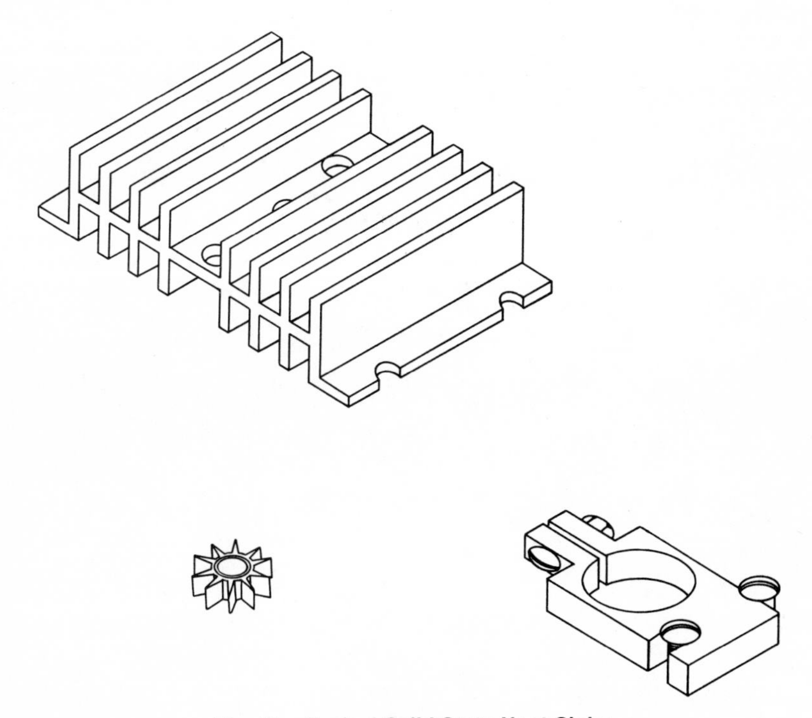

Fig. 12. Typical Solid State Heat Sinks

Each diode (rectifier) is rated as to its maximum PN junction temperature rating. Design of the heat sink should provide enough cooling so the junction temperature never reaches this maximum. In actuality it is recommended that temperatures never reach more than half the maximum.

The other factor in selecting a heat sink is its thermal resistance. Thermal resistance is calculated in terms of degrees centigrade divided by watts (or °C/W). Heat is transferred from the junction to the rectifier case and is designated θja or θjc by the manufacturer's specifications. From the case, heat moves into the heat sink. This heat transfer or path from case to heat sink is designated θcs. Finally, the heat transfer or path from heat sink to ambient air is designated θsa. The total thermal resistance is the total of these factors or $R_{THERMAL} = \theta ja \pm \theta cs \pm \theta sa$. Manufacturers' specifications provide the thermal resistance of their heat sinks. They should be accepted rather than attempting to calculate the thermal resistance by oneself.

Between the rectifier and the heat sink, an insulator made of mica or anodized aluminum should be installed. Heat sinks should be chosen by their cooling capability and the size or shape of the container that is required to contain them.

Section 6
Circuit Analysis of Regulator Circuits

The most basic form of regulator is the zener diode regulator, mainly because it is simple to understand and use. Since its conception it has been utilized alone and combined with other more complex regulator circuits. It has become the workhorse of the solid state power supply field of design and is used extensively in this book.

Our explanation will begin with the zener diode regulator, then expand through more complex power supply regulators.

ZENER DIODE REGULATOR CIRCUIT (SEE FIG. 13.)

The zener diode is closely related to the diode rectifier. Its purpose, however, is to act as a voltage regulator, either by itself or with a transistor-regulated power supply. The reason for its regulatory ability is its unique construction.

The zener diode is constructed to operate in reverse of the normal conditions present in rectifier diodes. This reverse condition is made possible by carefully planned doping conditions when the diode is manufactured. The zener diode is basically a breakdown diode. It is installed so its bias is the reverse of the diode rectifier.

In Fig. 13(A) the current-voltage characteristic curve of a zener diode is shown. This particular curve shows normal diode operation with the zener diode forward biased. In the reverse condition the point at which the diode breaks down is called the *zener breakdown voltage* or the *avalanche breakdown*. As the voltage against the diode barrier approaches this zener point, current carrying electrons tunnel through the barrier, causing the zener effect. At the same time other electrons accelerate, knocking others loose from their valence rings in a process called *carrier multiplication*, causing the effect called *avalanche*. At that time the diode is said to be at *breakdown*. At this voltage point the current may increase greatly while the voltage remains the same. It is this reason that makes the breakdown diode a perfect voltage regulator.

In actuality, the zener point and the avalanche point occur at different levels. The avalanche occurs slightly after the "knee" of

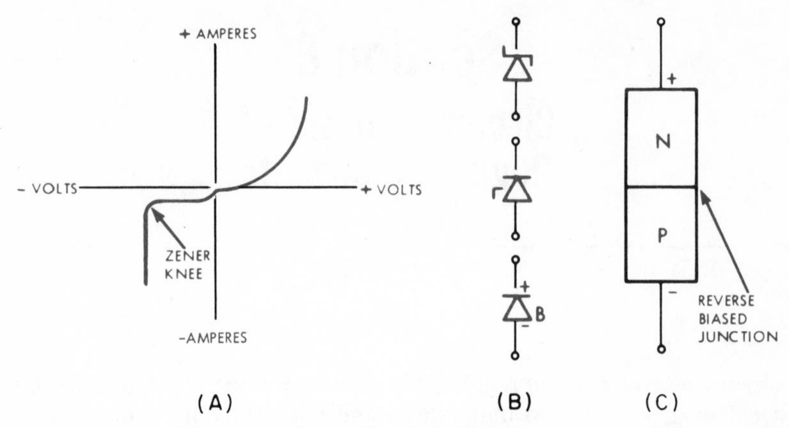

Fig. 13. Zener Diode, Showing: (A) Current-Voltage Characteristic Curve; (B) Zener and Breakdown Diode Symbols; and (C) Functional Diagram of Zener Diode

the Zener. For all practical purposes, the two effects can be said to happen together. Fig. 13(B) illustrates typical zener and breakdown diode schematic symbols. Fig. 13(C) is a functional diagram of the zener diode. A zener diode is built into a package similar to normal rectifiers.

DUAL OUTPUT ZENER DIODE REGULATOR CIRCUIT (SEE FIG. 14.)

The zener diode is sometimes used alone for voltage regulation. When using zener diodes by themselves for voltage regulation, a limiting resistor R_S is always placed in series with the load and the shunted zeners. In this way the unregulated output is protected from overload current.

The power rating of the limiting resistor R_S must be very high because it carries the load current and the current through the zeners. This makes the efficiency of the shunted zener regulator very low. Its simplicity makes up for its inefficiency. The zener diode is chosen for power-carrying capacity and its voltage setting. Large current zeners require heat sinks.

The figure shows the limiting resistance R_S in series with two zener diodes CR_1 and CR_2 and two loads. The zeners are regulating two load voltages at set levels. If the unregulated dc supply is large enough, it is possible to place several zeners in series with each other and provide multiple pickoffs similar to the ones illustrated.

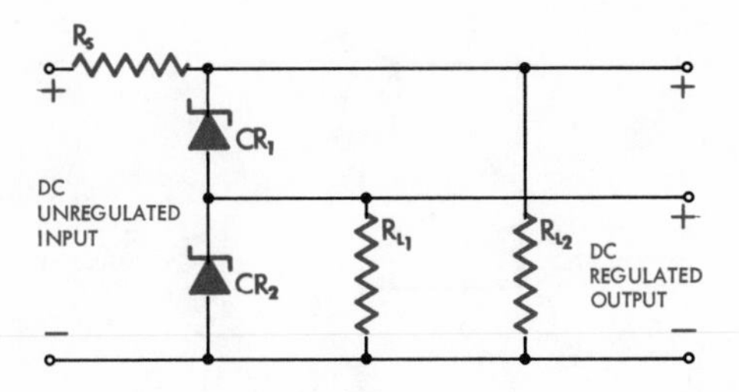

ANALYSIS:

1 CR1 and CR2 are voltage regulator zener diodes in series with each other and in parallel with the load.

2 Any change in load voltage will be felt on RL. Voltage across zener diodes always remains the same.

3 R_S provides short protection for the unregulated input.

4 Load no. 1 (RL1) is picked off center tap between zener diodes.

5 Load no. 2 (RL2) is across both zener diodes.

Fig. 14. Dual-Output Zener Diode Voltage Regulator

SERIES VOLTAGE REGULATOR CIRCUIT (SEE FIG. 15.)

The series regulator consists basically of a power transistor Q_1, a current-limiting resistor R_1, and a zener (voltage regulator) diode CR_1. The transistor used in the illustration is a PNP power transistor. It is chosen for its current-carrying capabilities. The power transistor must be placed on a heat sink when installed in the power regulator circuit. The heat sink is made of soft metal which absorbs heat. The zener diode on the base of the regulator, along with the load voltage, provides the base-emitter bias to control the transistor's operation. Note that the transistor is placed in the circuit in series with the load.

Bias for the transistor Q_1 is provided by the load voltage and the zener diode CR_1. If the voltage increases across the load, the bias current on the transistor decreases, current through the transistor decreases, therefore the voltage-drop across the transistor increases, causing the load voltage to lower back to normal.

In the event of a decrease in load voltage, bias current on the transistor Q_1 increases, current through the transistor increases, therefore the voltage drop across the transistor decreases, causing the load voltage to rise back to normal.

It must be noted that all these events take place simultaneously, and the moment a change is felt, regulation takes place. If you were

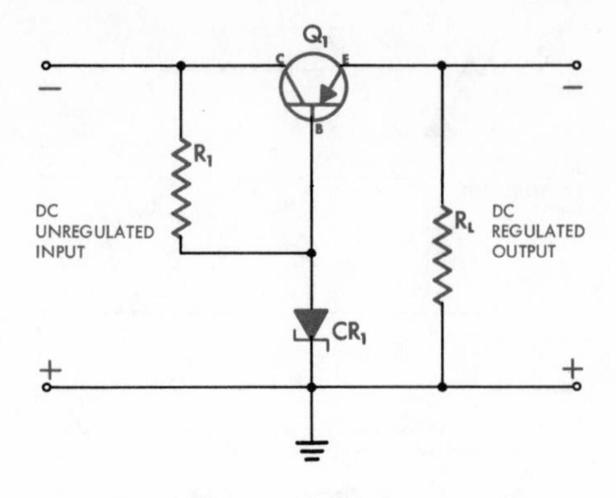

ANALYSIS:

1 CR1 and load voltage provide bias to Q1.

2 R1 limits current for the fixed voltage zener doide.

3 If load voltage increases, bias current on Q1 decreases, current through Q1 decreases, voltage drop across Q1 increases, and load voltage lowers to normal.

4 If load voltage decreases, bias current on Q1 increases, current through Q1 increases, voltage drop across Q1 decreases and load voltage rises to normal.

Fig. 15. Series Voltage Regulator Circuit

to put a meter across the load it would be difficult to see regulation happening.

SHUNT VOLTAGE REGULATOR CIRCUIT (SEE FIG. 16.)

The shunt regulator consists of a current-limiting resistor R_1, a zener (voltage regulator) diode CR_1 and a power transistor Q_1. The transistor used in the illustration is a PNP power transistor. It is chosen for its current-carrying capabilities. The power transistor here must be placed on a heat sink as in the series regulator circuit. The zener diode is installed in the base circuit, to provide a reference for bias. Bias for base-emitter junction is provided by the load voltage and current-limiting resistor R_1. Note that the transistor is placed in the circuit in shunt (parallel) with the load.

If the voltage across the load increases, the bias current increases, current through the transistor increases, and the voltage drop across the transistor decreases, causing the load voltage to lower back to normal.

In the event of a decrease in load voltage, the bias decreases,

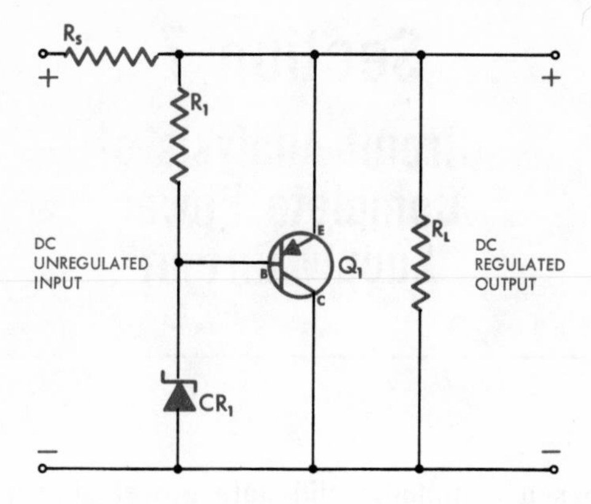

ANALYSIS:

1 R1 and load voltage provide bias to Q1.

2 CR1 is a fixed voltage zener diode.

3 If load voltage increases, voltage drop across R1 increases, bias current increases, current through Q1 increases, voltage drop across Q1 decreases, and load voltage lowers to normal.

4 If load voltage decreases, voltage drop across R1 decreases, bias current decreases, current through Q1 decreases, voltage drop across Q1 increases, and load voltage rises to normal.

5 R_s provides short protection for the unregulated input.

Fig. 16. Shunt Voltage Regulator Circuit

current through the transistor decreases, therefore the voltage drop across the transistor increases, causing the load voltage to rise to normal.

It must be noted that all these events take place simultaneously, and the moment a change is felt, regulation takes place.

CONTROL OF LOADING FOR POWER SUPPLIES

In the designs presented, regulators are used at the supply output to give constant voltage for approximately 30% load variations. This is desireable to prevent output voltage fluctuations due to load changes or input voltage variations.

Additionally, the regulators serve to reduce ripple to a negligible level. This is possible because the emitter of the regulation transistor follows the base, which is zener-controlled. Therefore voltage fluctuations at the collector are not seen at the emitter. If a power supply is designed without a regulator at the output, the ripple level and voltage output will vary in proportion to the load applied.

Section 7

Circuit Analysis of Complete Power Supply Circuit

Fig. 17 shows a complete solid-state power supply. It consists, from left to right, of a transformer T_1, current-limiting resistor R_S, diode rectifiers CR_1 and CR_2, the filter $C_1 - L_1 - C_2$, voltage-divider resistor R_1, and the power transistor Q_1. The resistance R_L represents the load.

As an analysis, unregulated AC power is fed from the transformer through a full-wave rectifier. It is then filtered by the *pi* filter, then regulated by a series regulator. The load is therefore dc regulated.

Test points *A* through *G* are shown on this circuit to show the reader where he would monitor the different devices in a trouble-shooting situation.

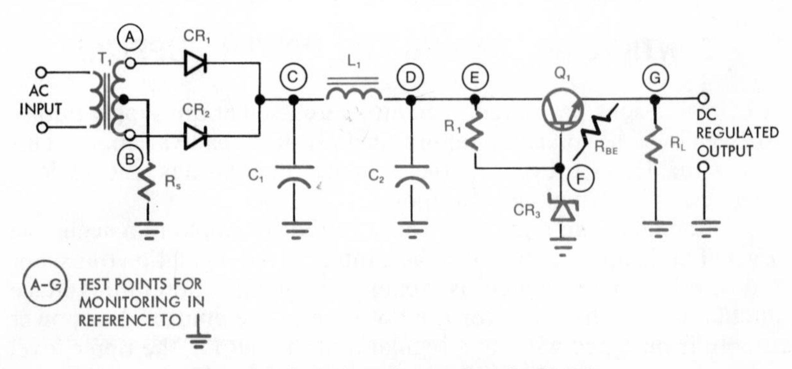

Fig. 17. Complete Power Supply Circuit

Section 8
Choosing Power Supply Components

The choice of power supply components is one that must be left up to the designer. Many components are available and the choices are not really as much a science as an art. One must choose what is available, and choices are usually concerned with cost.

Whatever your personal requirements, the basic guidelines are set forth in the next several paragraphs.

CHOOSING A TRANSFORMER

Power transformers are cataloged by manufacturers' specifications as to their volt-ampere (va) ratings. Since power in the primary of the transformer is equal to power in the secondary of the transformer, determination of primary and secondary voltages are selective. The only concern in the primary voltage is that the transformer primary matches the available source ac voltage. The secondary ac voltage in power supply transformers varies under standard values from 6 to 30 vac (and more in some cases).

The dc voltage that can be derived from a power transformer is from 1.25 to 1.4 times the secondary rms voltage. This value will vary depending on the type of power supply and the method of filtering. The power rating of the secondary should be 1.25 to 1.4 times the dc voltage out of the rectifiers. In some cases it may be desireable to have a larger power (va) rating.

CHOOSING RECTIFIER DIODES (REFER TO APPENDIX B)

Diode specifications are written for the forward-biased diode and reverse-bias conditions. Forward characteristics that are specified are forward voltage drop and forward current. Reverse characteristics are maximum reverse voltage (peak inverse voltage) and reverse current.

Forward voltage drop is the amount of voltage that is dropped under a normal forward-bias condition. This voltage drop will vary up to around 1.0 volt.

Forward current is the maximum current that can flow in the

diode without damaging the rectifier. Normal diode forward current would be about twice the output dc current.

Maximum reverse voltage (peak inverse voltage) rating is the amount of reverse-bias (voltage) the diode can withstand before it breaks down. In design, a diode (rectifier) should be chosen that withstands twice the output dc voltage of the power supply.

Reverse current (leakage current) is a small amount of current that flows in reverse when the diode is reverse-biased. This current should be as small as possible and, for the purpose of design, less than 1 milliampere.

CHOOSING A CAPACITIVE INPUT FILTER

The capacitive input filter is the basic filter used in solid-state power supplies. In most cases it is the only type filter required. It is always used in circuits where good regulation is not a requirement and in high voltage applications.

Capacitive inputs are recommended over choke input filtering because of simplicity. Values of capacitance are dependent on load current, ripple frequency, and ripple percentage required.

Load Current. Load current is the direct current (dc) that flows through the load. This current should not be misconstrued to be effective ac voltage (measured with an ac meter.) Furthermore, it should not be related to peak current through power supply rectifiers. The load current is the average value of dc current flowing through the load. Load current and load voltage are necessary to calculate load resistance (R_L) which, in turn, is used to calculate filter capacitance values. Load resistance formula is as follows:

$$R_{LOAD} = \frac{V_{DC\ LOAD}}{I_{DC\ LOAD}}, \text{ or } R_L = \frac{V_{DC}}{I_{DC}}$$

Ripple Frequency. The ripple frequency of a dc power supply for half-wave rectification is the same as the cycles per second (hertz). That is, a half wave 60 hertz power supply would have a 60 hertz ripple frequency. Ripple frequency for a full wave rectifier power supply would have a ripple frequency equal to the total negative and positive alternations. That is, a full-wave 60 hertz power supply would have a 120 hertz ripple frequency. Ripple frequency is represented by the symbol f_r.

Ripple Percentage. Output ripple is measured or calculated in percent of ripple. This output ripple may vary greatly from near 100% in half-wave rectifiers to as low as 2% or 3% in three to six-phase power-supply connections. Ripple percentage (r_p) is the ratio of the rms ripple voltage at the rectifier output to the dc voltage across the load. Ripple percent is calculated using the following formula:

$$\text{Ripple percent } (r_p) = \frac{V_{rms}}{V_{DC}} \times 100$$

Choosing a Capacitive Input Filter for a Half-Wave-Rectifier Power Supply. One basic formula for choosing a capacitive input filter for a single (half) wave rectifier power supply is as follows:

$$C = \frac{\sqrt{2}}{2\pi f_r R_L r_p}$$

WHERE:

C = Capacitance in farads
π = Constant 3.1416
f_r = Ripple frequency
R_L = Load resistance $\frac{V_{DC}}{I_{DC}}$
r_p = Ripple percentage (1% = 0.01)

EXAMPLE:

V_{DC} = 15 volts
I_{DC} = 1 amp
R_L = 15 ohms
f_r = 60 hertz
r_p = 1% or 0.01

$$C = \frac{1.414}{2 \times 3.1416 \times 60 \times 15 \times 0.01}$$

$$C = \frac{1.414}{56.55}$$

$$C = 25{,}000\ \mu F \text{ (approx.)}$$

In the following table is a list of capacitance values for a 15 vdc, 1 ampere power supply with various ripple percentages.

RIPPLE %	*CAPACITANCE* (in μF)
1%	25,000
2%	12,500
3%	8,333
4%	6,250
5%	5,000
10%	2,500

Note: *These are approximate values. Availability of standard sizes will prevail.*

Choosing a Capacitive Input Filter for a Full-Wave Rectifier Power Supply. Calculations for full-wave rectifier power supplies are the same as half-wave with the exception of the change in ripple frequency. Since a full wave rectifier uses both negative and positive alternations, the ripple voltage of a full wave rectifier is twice that of a half wave rectifier. Calculations are as follows:

$$C = \frac{\sqrt{2}}{2\pi f_r R_L r_p}$$

WHERE:

C = Capacitance in farads
π = Constant 3.1416
f_r = Ripple frequency
R_L = Load resistance $\frac{V_{DC}}{I_{DC}}$
r_p = Ripple percentage (1% = 0.01)

EXAMPLE:

$V_{DC} = 15$ volts
$I_{DC} = 1$ amp
$R_L = 15$ ohms
$f_r = 120$ hertz
(twice the half wave)
$r_p = 1\%$ or 0.01

$$C = \frac{1.414}{2 \times 3.1416 \times 120 \times 15 \times 0.01}$$

$$C = \frac{1.414}{113.1}$$

$$C = 12{,}500\ \mu F \text{ (approx.)}$$

In the following table is a list of capacitance values for a 15 vdc, 1 ampere power supply with various ripple percentages.

RIPPLE %	*CAPACITANCE* (in μF)
1%	12,500
2%	6,250
3%	4,166
4%	3,125
5%	2,500
10%	1,250

Note: *These are approximate values. Availability of standard sizes will prevail.*

CHOOSING A CHOKE INPUT FILTER

The choke input filter is very effective in reducing ripple in power supplies designed for high-current applications. The penalty, of course, is weight, which is usually objectionable.

The choke input filter is selected on the basis of the amount of current to be passed and the inductance (henrys) required to filter the power-supply ripple to an acceptable level. Typically, for most filter applications, 50 to 100 millihenries of inductance will provide satisfactory filtering. The limitation on the current passed is the point where the choke core becomes saturated and the ac impedance is reduced to the dc resistance of the winding. Additionally, winding wire size should be considered for proper rating. To ensure that the filter is operated within limits, the manufacturer's specifications should be consulted.

CHOOSING A ZENER DIODE REGULATOR

The basic demand of a zener diode regulator is to choose a zener value equal to the regulated voltage need of the load. Other requirements are power and current capacities.

Zener voltages range from very small, say 1.0 volt, to very large, say 28 volts. The zener voltage is the avalanche voltage at which the diode breaks down in a reverse-bias situation.

Depending on the severity of load changes, the zener diode will regulate voltage to the load at the rated value ±10% to 20%. If the voltage in reverse bias stays below the breakdown point, the zener diode will not go into avalanche; therefore no regulation will be present. Furthermore, if the reverse-bias voltage lowers to a point

25% or so below the zener point, regulation will not be maintained. (This point varies with the zener diode and design application.)

Zener diodes normally have low current ratings, under 250 milliamps, although higher current ratings are available. Proper size heat sinks are required with high-current rated zeners.

Power capacity is sometimes given as maximum power dissipation. The maximum power rating is derived by multiplying the zener voltage by the maximum current rating.

CHOOSING A REGULATOR POWER TRANSISTOR

A transistor for a power supply regulator is chosen in a similar manner to any circuit which has specific current and/or power requirements. The transistor must be able to withstand at least twice the peak power load that may be encountered. Most power supplies that have regulators are large enough so that they also must have heat sinks on which to mount the transistor.

If the transistor used is made of silicon, the voltage drop of the base-emitter junction is around 0.6 vdc. If the transistor used is made of germanium, the voltage drop is around 0.2 vdc. This, of course, is simply a rule of thumb. Each manufacturer provides curves that depict this base-emitter voltage drop at different temperatures. However, in the case of power supplies the curves do not normally have such exact requirements.

The gain (beta) of the transistor used should be from 50 to 100. This is also very flexible, and most power transistors have the ability to do the job.

CHOOSING A HEAT SINK

As previously described, heat sinks are rather complex components to deal with. Basically the heat sink must be able to dissipate enough heat to prevent the transistor or the diodes from burning up. Power transistors and diodes are usually installed with recommended heat sinks. There are nomographs for choosing the heat sinks. These are also in transistor/diode specifications. Since this is a complex determination, it is recommended that heat sinks be used as directed by these specifications. This practice will ensure a tested and tried design.

If expense is a problem, choose a finned heat sink as large as can be used within reason. The area of the heat sink, along with the material the sink is made from, are the basic factors to be considered. Again, on-shelf, tested and tried heat sinks are recommended.

Section 9

Designing a Half-Wave Rectifier Power Supply

The half-wave rectifier power supply is the simplest of all the power supplies. It is also the least efficient. The reason for this is that only half the input sine wave becomes dc and the other half is blocked by reverse bias of the diode.

The half-wave rectifier power supply is used where expense must be kept to a minimum and where good rectification is not a necessity. The half-wave rectifier has few parts: the transformer, a rectifier diode, a voltage-divider resistor, a filter capacitor, a power transistor, and a zener diode regulator. Fig. 18 is a circuit diagram for this type of power supply.

GLOSSARY FOR HALF-WAVE RECTIFIER POWER SUPPLY DESIGN (SEE FIG. 18.)

C_1 Shunt capacitor
CR_1 Rectifier number 1
f_r Ripple frequency
I_B Transistor base current
I_{DC} Zener diode current plus load current
I_{R_L} Load current
I_{R_1} Current through resistor R_1
I_Z Zener diode current
P_D Power dissipation of transistor Q_1
P_{LOAD} Power through load
P_{B_1} Power rating of resistor R_1
P_{SEC} Power rating of T_1 transformer secondary
Q_1 Power transistor
R_1 Resistor number 1
R_{BE} Resistance base emitter junction
R_{TL} Total load resistance as seen by C_1
r_p Ripple percentage of unregulated dc voltage
V_{CR_2} Zener diode voltage
V_{DC} Unregulated dc voltage
V_{R_L} Load voltage
V_{PRI} Voltage of T_1 transformer primary
V_{R_1} Voltage drop across resistor R_1

$V_{R_{BE}}$ Voltage drop across base emitter junction
V_{SEC} Voltage of T_1 transformer secondary
T_1 Transformer

DESIGN PROCEDURE FOR A HALF-WAVE RECTIFIER POWER SUPPLY (FIG. 18.)

(1) Determine load voltage V_{R_L} and dc unregulated voltage V_{DC}. (Approximately 1.4 V_{R_L}) (GIVEN)

(2) Determine load current, I_{R_L}. (GIVEN)

(3) Calculate load power, $P_{LOAD} = V_{R_L} \times I_{R_L}$

(4) Choose power transistor Q_1 (with heat sink) to handle 2 × P_{LOAD} power dissipation P_D.

Note: *If* Q_1 *transistor is made of silicon, the base-emitter junction voltage drop* $V_{R_{BE}}$ *is approximately 0.6 vdc. If* Q_1 *transistor is made of germanium, the base emitter junction voltage drop is approximately 0.2 vdc.*

(5) Choose a base current for Q_1. $I_B = \frac{I_{R_L}}{\text{beta}}$

(6) Choose a zener diode current $I_Z = 2 \times I_B$.

(7) Calculate $I_{DC} = I_Z + I_{R_L}$

(8) Choose a zener diode voltage $V_{CR_2} = V_{R_L} + V_{R_{BE}}$

(9) Choose a zener diode to meet (6) and (8) requirements.

(10) Calculate $V_{R_1} = V_{DC} - V_{CR_2}$

(11) Calculate $I_{R_1} = I_Z + I_B$

(12) Calculate resistor $R_1 = \frac{V_{R_1}}{I_{R_1}}$

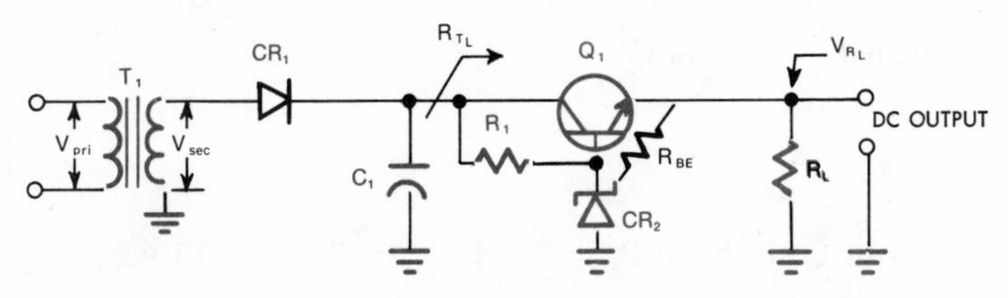

Fig. 18. Half-Wave Rectifier Design Schematic

(13) Calculate resistor R_1 power rating $P_{R_1} = V_{R_1} \times I_{R_1}$

(14) Choose resistor R_1 to meet (12) and (13) requirements

(15) Calculate total load resistance $R_{TL} = \frac{V_{DC}}{I_{DC}}$

(16) Calculate $C_1 = \frac{\sqrt{2}}{2\pi f_r R_{TL} r_p}$

(17) Calculate C_1 working voltage $= 2 \times V_{DC}$

(18) Choose capacitor C_1 to have (16) and (17) requirements

(19) Choose rectifier CR_1. (See manufacturers' specifications.)
- (a) Reverse voltage rating must be approximately $2 \times V_{DC}$.
- (b) Leakage current (reverse current rating) must be less than 1 milliamp.
- (c) Forward current rating must be greater than $2 \times I_{DC}$.

(20) Choose power transformer T_1. (See manufacturers' specifications.)
- (a) Ensure transformer is 60 or 400 cycle as required.
- (b) Select primary volts (V_{PRI}) as required (110 vac or 220 vac)
- (c) Calculate secondary volts (V_{SEC}) as required. $V_{SEC} = \frac{V_{DC}}{1.3}$
- (d) Calculate secondary power $P_{SEC} = 0.65 \times V_{DC}$
- (e) Ensure that secondary dc resistance is greater than $3 \times V_{DC}$.
- (f) Select T_1 transformer to meet steps (a) through (e) requirements.

DESIGN EXAMPLE FOR HALF-WAVE RECTIFIER POWER SUPPLY (Fig. 18.)

(1) Given $V_{R_L} = 18$ vdc, $V_{DC} = 1.4 \times 18 = 25.2$ vdc

(2) Given $I_{R_L} = 0.5$ amp

(3) $P_{LOAD} = V_{R_L} \times I_{R_L}$
$P_{LOAD} = 18 \times 0.5$
$P_{LOAD} = 9$ watts

(4) Choose Q_1 to have $P_D = 2 \times P_{LOAD} = 2 \times 9 = 18$ watts

(5) Calculate Q_1 base current

$$I_B = \frac{I_{R_L}}{beta}$$

$$I_B = \frac{0.5}{50}$$

$$I_B = 0.01 \text{ amp}$$

(6) $I_Z = 2 \times I_B = 2 \times 0.01 = 0.02$ amp

(7) $I_{DC} = I_Z + I_{R_L} = 0.02 + 0.50 = 0.52$ amp

(8) $V_{CR_2} = V_{R_L} + V_{R_{BE}} = 18 + 0.6 = 18.6$ vdc

(9) Choose a zener diode to meet (6) and (8) requirements.

(10) $V_{R_1} = V_{DC} - V_{CR_2} = 25.2 - 18.6 = 6.6$ vdc

(11) $I_{R_1} = I_Z + I_B = 0.02 + 0.01 = 0.03$ amp

(12) $R_1 = \frac{V_{R_1}}{I_{R_1}} = \frac{6.6}{0.03} = 220$ ohms

(13) $P_{R_1} = V_{R_1} \times I_{R_1} = 6.6 \times 0.03 = 0.198$ watt

(14) Choose R_1 to meet (12) and (13) requirements (use $\frac{1}{2}$ watt standard)

(15) $R_{TL} = \frac{V_{DC}}{I_{DC}} = \frac{25.2}{0.52} = 48.5$ ohms

(16) $C_1 = \frac{\sqrt{2}}{2\pi f_r R_L r_p}$

GIVEN:
$f_r = 60$ Hz (half wave)
$r_p = 5\%$, or 0.05

$$C_1 = \frac{1.414}{2 \times 3.1416 \times 60 \times 48.5 \times 0.05}$$

$$C_1 = \frac{1.414}{914.206}$$

$$C_1 = 0.001546 = 1546\ \mu F$$

(17) C_1 working volts $= 2 \times 25.2 = 50.4$ volts

(18) Choose C_1 to meet (16) and (17) requirements

(19) Choose rectifier C_{R1} by specification

(20) Choose transformer T_1
(a) 60 hertz
(b) 110 vac

(c) $\frac{V_{DC}}{1.3} = \frac{25.2}{1.3} = 19.38$ vdc (This is approximate and transformer should be chosen as close to this value as practical.)

(d) $P_{SEC} = 0.65 \times V_{DC} = 0.65 \times 25.2 = 16.38$ watts

(e) T_1 − dc resistance $= 3 \times V_{DC} = 3 \times 25.2 = 75.6$ ohms

(f) Choose T_1 to satisfy (a) through (e)

Section 10

Designing A Full-Wave Rectifier Power Supply

The full-wave rectifier is used with a center-tap transformer. It is, along with the bridge rectifier, the most used of the power supply types. The full-wave rectifier uses both alternations of the ac sine wave input and provides efficient dc output.

The full-wave rectifier power supply consists of a center-tap transformer, a pair of rectifier diodes, a filter capacitor, a voltage-divider resistor, a power transistor, and a zener diode regulator. Fig. 19 is a circuit diagram for this type of power supply.

GLOSSARY FOR FULL-WAVE RECTIFIER POWER SUPPLY DESIGN (SEE FIG. 19.)

C_1 Shunt capacitor
CR_1 Rectifier number 1
CR_2 Rectifier number 2
f_r Ripple frequency ($2 \times$ source hertz)
I_B Transistor base current
I_{DC} Zener diode current plus load current
I_{R_L} Load current
I_{R_1} Current through resistor R_1
I_Z Zener diode current
P_D Power dissipation of transistor Q_1
P_{LOAD} Power through load
P_{R_1} Power rating of resistor R_1
P_{SEC} Power rating of T_1 transformer secondary
Q_1 Power transistor
R_1 Resistor number 1
R_{BE} Resistance base emitter junction
R_L Load resistance
r_p Ripple percentage of unregulated dc voltage
R_{TL} Total load resistance as seen by C_1
V_{CR_3} Zener diode voltage
V_{DC} Unregulated dc voltage
V_{R_L} Load voltage
V_{PRI} Voltage of T_1 transformer primary

V_{R_1}	Voltage drop across resistor R_1
$V_{R_{BE}}$	Voltage drop across base-emitter junction
V_{SEC}	Voltage of T_1 transformer secondary
T_1	Transformer

DESIGN PROCEDURE FOR FULL-WAVE RECTIFIER POWER SUPPLY (FIG. 19.)

① Determine load voltage V_{R_L} and dc unregulated voltage V_{DC}. (Approximately 1.4 V_{R_L}) (GIVEN)

② Determine load current, I_{R_L} (GIVEN)

③ Calculate load power, $P_{LOAD} = V_{R_L} \times I_{R_L}$

④ Choose power transistor Q_1 (with heat sink) to handle 2 × P_{LOAD}

Note: *If Q_1 transistor is made of silicon, the base-emitter junction voltage drop $V_{R_{BE}}$ is approximately 0.6 vdc. If Q_1 transistor is made of germanium, the base emitter junction voltage drop is approximately 0.2 vdc.*

⑤ Choose a base current for Q_1. $I_B = \frac{I_{R_L}}{\text{beta}}$

⑥ Choose a zener diode current $I_Z = 2 \times I_B$.

⑦ Calculate I_{DC} ... $I_Z + I_{R_L}$

⑧ Choose a zener diode voltage $V_{CR_3} = V_{R_L} + V_{R_{BE}}$

⑨ Choose a zener diode to meet ⑥ and ⑧ requirements.

⑩ Calculate $V_{R_1} = V_{DC} - V_{CR_3}$

⑪ Calculate $I_{R_1} = I_Z + I_B$

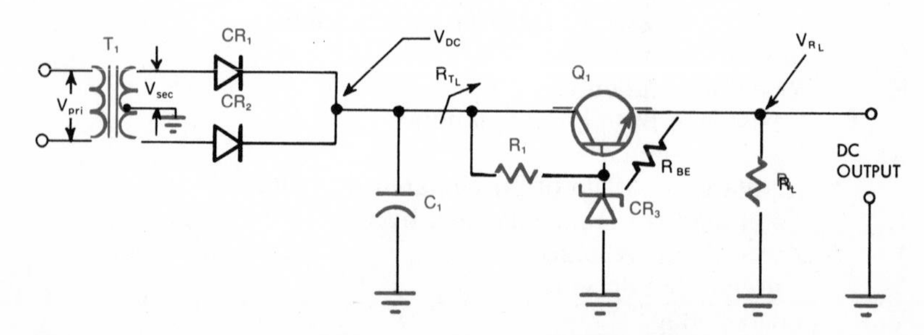

Fig. 19. Full-Wave Rectifier Design Schematic

(12) Calculate resistor $R_1 = \frac{V_{R_1}}{I_{R_1}}$

(13) Calculate resistor R_1 power rating $P_{R_1} = V_{R_1} \times I_{R_1}$

(14) Choose resistor R_1 to have (12) and (13) requirements

(15) Calculate total load resistance $R_{TL} = \frac{V_{DC}}{I_{DC}}$

(16) Calculate $C_1 = \frac{\sqrt{2}}{2\pi f_r R_{TL} r_p}$

(17) Calculate c_1 working voltage $= 2 \times V_{DC}$

(18) Choose capacitor c_1 to have (16) and (17) requirements

(19) Choose rectifiers CR_1 and CR_2. (See manufacturers' specifications.)
(a) Reverse voltage ratings must be greater than $2 \times V_{DC}$.
(b) Leakage current (reverse current rating) must be less than 1 milliamp.
(c) Forward current rating must be greater than $2 \times I_{DC}$.

(20) Choose power transformer T_1. (See manufacturers' specifications.)
(a) Ensure transformer is 60 or 400 cycle as required.
(b) Select primary volts (V_{PRI}) as required (110 vac or 220 vac)
(c) Calculate secondary volts (V_{SEC}) as required. $V_{SEC} = \frac{V_{DC}}{1.3}$
(d) Calculate secondary power $P_{SEC} = 1.3 \times V_{DC}$
(e) Ensure that secondary dc resistance is greater than $3 \times V_{DC}$
(f) Select T_1 transformer to meet (a) through (e) requirements.

DESIGN EXAMPLE FOR FULL-WAVE RECTIFIER POWER SUPPLY (FIG. 19.)

(1) Given $V_{DC} = 20$ vdc and $V_{R_L} = 15$ vdc

(2) Given $I_{R_L} = 1$ amp

(3) $P_{LOAD} = V_{R_L} \times I_{R_L}$
$P_{LOAD} = 15 \times 1$
$P_{LOAD} = 15$ watts

(4) Choose Q_1 to have $P_D = 2 \times P_{LOAD} = 2 \times 15 = 30$ watts

(5) $I_B = \frac{I_{R_L}}{beta}$. Let beta = 50 to 100 (whatever is the beta of the transistor chosen)

$I_B = \frac{1}{50}$

$I_B = 0.020$ amp

(6) $I_Z = 2 \times I_B = 2 \times 0.020 = 0.040$ amp

(7) $I_{DC} = I_Z + I_{R_L} = 0.04 + 1.0 = 1.04$ amps

(8) $V_{CR_3} = V_{R_L} + V_{R_{BE}} = 15 + 0.6 = 15.6$ vdc

(9) Choose a zener diode to meet (6) and (8) requirements.

(10) $V_{R_1} = V_{DC} - V_{CR_3} = 20 - 15.6 = 4.4$ vdc

(11) $I_{R_1} = I_Z + I_B = 0.040 + 0.20 = 0.060$ amp

(12) $R_1 = \frac{V_{R_1}}{I_{R_1}} = \frac{4.4}{0.060} = 73.3$ ohms

(13) $P_{R_1} = V_{R_1} \times I_{R_1} = 4.4 \times 0.060 = 0.264$ watt

(14) Choose resistor R_1 to meet (12) and (13) requirements.

(15) $R_{TL} = \frac{V_{DC}}{I_{DC}} = \frac{20}{1.04} = 19.23$ ohms

(16) $C_1 = \frac{\sqrt{2}}{2\pi f_r R_{TL} r_p}$

GIVEN:
$f_r = 120$ (60 hertz full wave)
$r_p = 5\%$, or 0.05 (ripple percent)

$C_1 = \frac{1.414}{2 \times 3.1416 \times 120 \times 19.23 \times 0.05}$

$C_1 = \frac{1.414}{724.956}$

$C_1 = 0.001950$ or 1950 μF

(17) C_1 working voltage $= 2 \times V_{DC} = 40$ volts

(18) Choose capacitor c_1 to meet (16) and (17) requirements.

(19) Choose rectifiers CR_1 and CR_2.

(20) Choose transformer T_1
(a) 60 hertz
(b) 110 vac
(c) $V_{SEC} = \frac{V_{DC}}{1.3} = \frac{20}{1.3} = 15.38$ vdc (This is approximate and should be chosen as close to this value as practical)

(d) $P_{SEC} = 1.3 \times V_{DC} = 1.3 \times 20 = 26$ watts
(e) T_1 – dc resistance $= 3 \times V_{DC} = 3 \times 20 = 60$ ohms
(f) Choose T_1 to satisfy step (a) through (e).

Section 11
Designing A Bridge Rectifier Power Supply

The bridge rectifier is used without a center-tapped transformer. It is, along with the full wave rectifier, the most used of the power supply types. The bridge rectifier uses both alternations of the ac sine wave input and provides efficient dc output. Additionally, the bridge rectifier withstands twice the reverse voltage that a full wave rectifier can.

The bridge rectifier consists of a transformer, a bridge of four rectifier diodes, a filter capacitor, a voltage-divider resistor, a power transistor, and a zener diode regulator. Fig. 20 is a schematic diagram of this circuit.

GLOSSARY FOR BRIDGE RECTIFIER POWER SUPPLY DESIGN (SEE FIG. 20.)

C_1	Shunt capacitor
CR_1	Rectifier number 1
CR_2	Rectifier number 2
CR_3	Rectifier number 3
CR_4	Rectifier number 4
CR_5	Rectifier number 5
f_r	Ripple frequency (2 × source hertz)
I_B	Transistor base current
I_{DC}	Zener diode current plus load current
I_{R_L}	Load current
I_{R_1}	Current through resistor R_1
I_Z	Zener diode current
P_D	Power dissipation of transistor Q_1
P_{LOAD}	Power through load
R_{P_1}	Power rating of resistor R_1
P_{SEC}	Power rating of T_1 transformer secondary
Q_1	Power transistor
R_1	Resistor number 1
R_{BE}	Resistance of base-emitter junction
R_L	Load resistance
r_p	Ripple percentage of unregulated dc voltage

R_{TL} Total load resistance as seen by C_1
V_{CR_5} Zener diode voltage rating
V_{DC} Unregulated dc voltage
V_{R_L} Load voltage
V_{PRI} Voltage of T_1 transformer primary
$V_{R_{BE}}$ Voltage drop across base emitter junction
V_{SEC} Voltage of T_1 transformer secondary
T_1 Transformer

DESIGN PROCEDURE FOR BRIDGE RECTIFIER POWER SUPPLY (FIG. 20.)

① Determine load voltage V_{R_L} and unregulated dc voltage V_{DC} (approximately 1.4 V_{R_L})

② Determine load current I_{R_L} (GIVEN)

③ Calculate load power, $P_{LOAD} = V_{R_L} \times I_{R_L}$.

④ Choose power transistor Q_1 (with heat sink) to handle $2 \times P_{LOAD}$ power dissipation (P_D)

Note: *If* Q_1 *transistor is made of silicon, the base-emitter junction voltage drop* $V_{R_{BE}}$ *is approximately 0.6 vdc. If* Q_1 *transistor is made of germanium, the base-emitter junction voltage drop is approximately 0.2 vdc.*

⑤ Choose a base current for Q_1. $I_B = \frac{I_{R_L}}{\text{beta}}$

⑥ Choose a zener diode current $I_Z = 2 \times I_B$.

⑦ Calculate $I_{DC} = I_Z + I_{R_L}$

⑧ Choose a zener diode voltage, $V_{CR_5} = V_{R_L} + V_{R_{BE}}$

⑨ Choose a zener diode to meet ⑥ and ⑧ requirements.

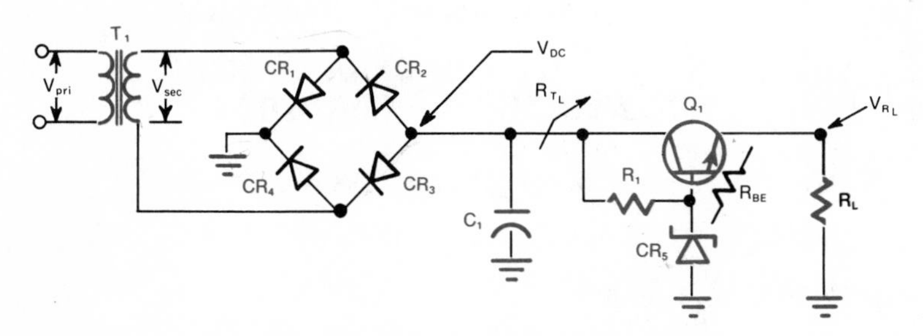

Fig. 20. Bridge Rectifier Design Schematic

(10) Calculate $V_{R_1} = V_{DC} - V_{CR_5}$

(11) Calculate $I_{R_1} = I_Z + I_B$

(12) Calculate $R_1 = \frac{V_{R_1}}{I_{R_1}}$

(13) Calculate R_1 power rating, $P_{R_1} = V_{R_1} \times I_{R_1}$

(14) Choose resistor R_1 to meet (12) and (13) requirements.

(15) Calculate $R_{TL} = \frac{V_{DC}}{I_{DC}}$

(16) Calculate $C_1 = \frac{\sqrt{2}}{2\pi f_r R_{TL} r_p}$

(17) Calculate C_1 working voltage $= 2 \times V_{DC}$

(18) Choose capacitor C_1 to meet (16) and (17) requirements.

(19) Choose rectifiers CR_1, CR_2, CR_3, and CR_4. (See manufacturers' specifications.)
(a) Reverse voltage ratings must be greater than V_{DC}.
(b) Leakage current (reverse current rating) must be less than 1 milliamp.
(c) Forward current rating must be greater than $2 \times I_{DC}$.

(20) Choose power transformer T_1. (See manufacturers' specifications).
(a) Ensure transformer is 60 or 400 hertz as required.
(b) Select primary volts (V_{PRI}) as required (110 vac or 220 vac).
(c) Calculate secondary volts (V_{SEC}) as required. $V_{SEC} = \frac{V_{DC}}{1.3}$
(d) Calculate secondary power $P_{SEC} = 1.3 \times V_{DC}$
(e) Ensure that secondary dc resistance is greater than $3 \times V_{DC}$
(f) Select transformer T_1 to meet (a) through (e) requirements.

DESIGN EXAMPLE FOR BRIDGE RECTIFIER POWER SUPPLY (FIG. 20.)

(1) Given $V_{R_L} = 18$ vdc, $V_{DC} = 1.4 \times 18 = 25.2$ vdc

(2) Given $I_{R_L} = 0.5$ amp

(3) $P_{LOAD} = V_{R_L} \times I_{R_L}$
$P_{LOAD} = 18 \times 0.5$
$P_{LOAD} = 9$ watts

(4) Choose Q_1 to have $P_D = 2 \times P_{LOAD} = 2 \times 9 = 18$ watts

(5) $Q_1 - I_B = \frac{I_{R_L}}{beta}$

$$I_B = \frac{0.5}{50}$$

$$I_B = 0.01 \text{ amp}$$

(6) $I_Z = 2 \times I_B = 2 \times 0.01 = 0.02$ amp

(7) $I_{DC} = I_Z + I_{R_L} = 0.02 + 0.5 = 0.52$ amp

(8) $V_{CR_5} = V_{R_L} + V_{R_{BE}} = 18 + 0.6 = 18.6$ vdc

(9) Choose a zener diode to meet (6) and (8) requirements

(10) $V_{R_1} = V_{DC} - V_{CR_5} = 25.2 - 18.6 = 6.6$ vdc

(11) $I_{R_1} = I_Z + I_B = 0.02 + 0.01 = 0.03$ amp

(12) $R_1 = \frac{V_{R_1}}{I_{R_1}} = \frac{6.6}{0.03} = 220$ ohms

(13) $P_{R_1} = V_{R_1} \times I_{R_1} = 6.6 \times 0.03 = 0.198$ watt (use $\frac{1}{2}$ watt standard)

(14) Choose R_1 to meet (12) and (13) requirements.

(15) $R_{L_T} = \frac{V_{DC}}{I_{DC}} = \frac{25.2}{0.52} = 48.5$ ohms

(16) $C_1 = \frac{\sqrt{2}}{2\pi f_r R_{L_T} r_p}$

GIVEN:
$f_r = 60$ (120 hertz full wave)
$r_p = 5\%$ or 0.05 (ripple frequency)

$$C_1 = \frac{1.414}{2 \times 3.1416 \times 120 \times 48.5 \times 0.05}$$

$$C_1 = \frac{1.414}{1828.411}$$

$C_1 = 0.0007733$F, or 773.3 μF

(17) C_1 working volts $= 2 \times 25.2 = 50.4$ volts

(18) Choose capacitor C_1 to meet (16) and (17) requirements.

(19) Choose rectifier CR_1, CR_2, CR_3, and CR_4 by specifications.

(a) Reverse voltage greater than $V_{DC} = 25.2$ volts
(b) Reverse current less than 1 milliamp
(c) Forward current $2 \times 0.52 = 1.04$ amps

(20) Choose transformer T_1:
(a) 60 hertz
(b) 110 vac
(c) $V_{SEC} = \frac{V_{DC}}{1.3} = \frac{25.2}{1.3} = 19.38$ vdc (This is approximate and should be chosen as close to this value as practical)
(d) $P_{SEC} = 1.3 \times 25.2 = 33.15$ watts
(e) DC resistance of $T_1 = 3 \times V_{DC} = 3 \times 25.2 = 75.6$ ohms
(f) Choose T_1 to satisfy (a) through (e)

Section 12

Designing A Zener-Diode Regulated Full-Wave Power Supply

The zener diode regulated full-wave rectifier power supply is the best supply to use when very low-power (therefore low-current) applications are desired.

With low current no heat sinks are required. Therefore the zener diode regulated power supply has the advantages of low weight and low costs. The disadvantages are poor-to-good filtering and limitations in current capacity.

The zener diode regulated power supply consists of a transformer, a pair of rectifier diodes, a filter capacitor, a surge resistor, and a zener diode regulator. Fig. 21 is a schematic diagram of this type of power supply.

GLOSSARY FOR ZENER-DIODE REGULATED, FULL-WAVE POWER SUPPLY (SEE FIG. 21.)

C_1	Shunt capacitor
CR_1	Rectifier number 1
CR_2	Rectifier number 2
CR_3	Zener diode
f_r	Ripple frequency
$I_{CR_3 max}$	Maximum allowable current through zener diode
$I_{CR_3 normal}$	Normal current through zener diode
I_{R_L}	Load current
I_{R_S}	Current through surge resistor R_S
P_{CR_3}	Zener diode power dissipation
P_{LOAD}	Power through load
P_{R_S}	Power rating of surge resistor
P_{SEC}	Power rating of T_1 transformer secondary
R_L	Load resistance
r_p	Ripple percentage of unregulated dc voltage
R_S	Surge resistor
R_{TL}	Total load resistance seen by C_1 capacitor
T_1	Transformer
V_{CR_3}	Zener diode voltage
$V_{DC_{in}}$	Unregulated dc voltage
V_{PRI}	Voltage of T_1 transformer primary

V_{R_L}	Load voltage
V_{SEC}	Voltage of T_1 transformer secondary
π	3.14 (approximately)
2π	6.28 (approximately)

DESIGN PROCEDURES FOR ZENER-DIODE REGULATED, FULL-WAVE POWER SUPPLY (FIG. 21.)

(1) Determine load voltage V_{R_L} (*GIVEN*)

Note: $V_{R_L} = V_{CR_3}$

(2) Determine load current I_{R_L} (*GIVEN*)

(3) Calculate load resistance $R_L = \frac{V_{R_L}}{I_{R_L}}$ (often given in lieu of I_{R_L})

(4) Calculate load power $P_{LOAD} = V_{R_L} \times I_{R_L}$ (often given in lieu of I_{R_L})

(5) Determine maximum $V_{DC_{in}} = 1.5 \times V_{CR_3}$.

(6) Calculate safe zener diode power dissipation $P_{CR_3} = 2 \times P_{LOAD}$

(7) Calculate maximum allowable zener diode current $I_{CR_3\,max} = \frac{P_{CR_3}}{V_{CR_3}}$

(8) Select zener diode per (1), (6) and (7) requirements.

Note: *Zener voltage* $V_{CR_3} = V_{R_{LOAD}}$

(9) Calculate normal zener diode current $I_{CR_3\,normal} = I_{R_L}$

(10) Calculate maximum $I_{R_S} = I_{R_L} + I_{CR_3\,normal}$

(11) Calculate $R_S = \frac{\max V_{DC_{in}} - V_{R_L}}{I_{R_S}}$

(12) Calculate R_S power rating $P_{R_S} = (V_{DC_{in}} - V_{R_L}) \times I_{R_S}$

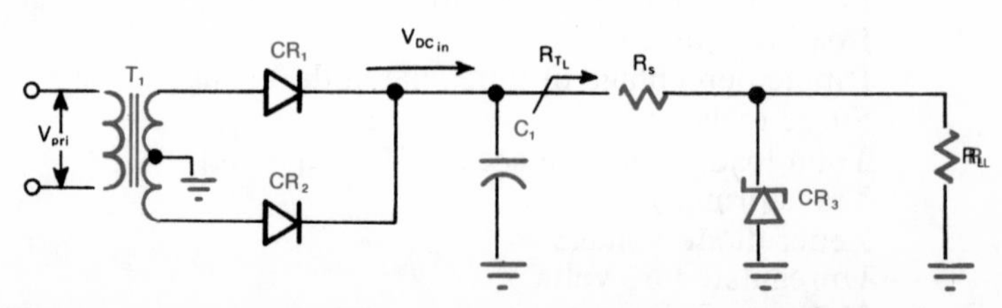

Fig. 21. Zener-Diode Regulated, Full-Wave Rectifier Design Schematic

(13) Select R_S per (10) and (11) requirements.

(14) Calculate $R_{TL} = R_L + R_S$

(15) Calculate capacitance value of $C_1 = \frac{\sqrt{2}}{2\pi f_r R_{TL} r_p}$

(16) Calculate capacitor C_1 working voltage $= 2 \times V_{DC_{in}}$

(17) Select capacitor C_1 per (13) and (14) requirements.

(18) Choose rectifiers CR_1 and CR_2 (see manufacturers specifications) as follows:
(a) Reverse voltage rating must be $2 \times V_{DC_{in}}$
(b) Leakage current (reverse current) rating must be less than 1 milliampere.

(19) Choose power transformer T_1 (See manufacturers specifications as follows:
(a) Ensure transformer is 60 or 400 hertz as required.
(b) Select primary volts (V_{PRI}) 110V or 220V as required.
(c) Calculate and select secondary volts (V_{SEC}) as required $V_{SEC} = \frac{V_{DC_{in}}}{1.3}$
(d) Calculate secondary power $P_{SEC} = 1.3 \times V_{DC_{in}}$

DESIGN EXAMPLE FOR ZENER-DIODE REGULATED, FULL-WAVE POWER SUPPLY (FIG. 21.)

(1) Given $V_{R_L} = 18$ vdc

(2) Given $I_{R_L} = 0.2$ amp

(3) $R_L = \frac{V_{R_L}}{I_{R_L}} = \frac{18}{0.2} = 90$ ohms

(4) $P_{LOAD} = V_{R_L} \times I_{R_L} = 18 \times 0.2 = 3.6$ watts

(5) Max $V_{DC_{in}} = 1.5 \times 18 = 27$ vdc

(6) $P_{CR_3} = 2 \times 3.6 = 7.2$ watts

(7) $I_{CR_3} = \frac{7.2}{18} = 0.4$ amp

(8) Select zener diode per (1), (6) and (7)

(9) $I_{CR_3 normal} = 0.2$ amp

(10) $I_{R_S} = 0.2 + 0.2 = 0.4$ amp

(11) $R_S = \dfrac{27 - 18}{0.4} = \dfrac{9}{0.4} = 22.5$ ohms

(12) $P_{R_S} = (27 - 18) \times 0.4 = 3.6$ watts

(13) Select R_S per (11) and (12)

(14) $R_{TL} = R_L + R_S = 90 + 22.5 = 112.5$ ohms

(15) $C_1 = \dfrac{\sqrt{2}}{2\pi f_r R_{T_L} r_p}$ *GIVEN:* $f_r = 120$ (60 hertz full wave)
$r_p = 5\%$ (ripple percent)

$$C_1 = \frac{1.414}{2 \times 3.1416 \times 120 \times 112.5 \times 0.05}$$

$$C_1 = \frac{1.414}{4241.16}$$

$C_1 = 0.0003333$ F, or 333.3 μF

(16) C_1 working voltage $= 2 \times 27 = 54$ vdc

(17) Select capacitor C_1 per (15) and (16)

(18) Choose rectifier CR_1 and CR_2 by specification:
(a) Reverse voltage rating $= 2 \times 27 = 54$ vdc
(b) Leakage current less than 1 milliamp

(19) Choose transformer T_1 by specification:
(a) 60 hertz
(b) 110 vac
(c) $V_{SEC} = \dfrac{27}{1.3} = 20.76$ vac
(d) $P_{SEC} = 1.3 \times 27 = 35.1$ watts

Section 13
Designing A Full-Wave Choke Input Filter Power Supply

The choke input filter provides the best filtering in a full-wave power supply. It is used when high-current (therefore high-power) applications are necessary.

The disadvantages of the choke input filter are excessive weight and higher costs. The advantages are better filtering and protection against surge current.

The choke input filter power supply consists of a transformer, a pair of rectifier diodes, a series choke, a filter capacitor, a divider resistor, a power transistor, and a zener diode regulator. Fig. 22 is a circuit diagram for this type of power supply.

GLOSSARY FOR FULL-WAVE CHOKE INPUT POWER SUPPLY (SEE FIG. 22.)

C_1	Shunt capacitor
CR_1	Rectifier number 1
CR_2	Rectifier number 2
CR_3	Zener diode
f_r	Ripple frequency
I_B	Transistor base current
I_{DC}	Zener diode plus load current
I_{R_1}	Resistor R_1 current
I_{R_L}	Load current
I_{CR_3}	Zener diode current
L_1	Series choke
P_D	Power dissipation of transistor Q_1
P_{LOAD}	Power through load
P_{R_1}	Power rating of resistor R_1
P_{SEC}	Power rating of T_1 transformer secondary
Q_1	Power transistor
R_1	Resistor number 1
R_{BE}	Resistance base emitter junction
r_{in}	Ripple percentage at L_1 input
r_{out}	Ripple percentage at L_1 output
r_p	Ripple percentage of unregulated dc voltage

R_{TL}	Total load resistance seen by C_1 capacitor
T_1	Transformer
V_{CR_3}	Zener diode voltage
V_{DC}	Unregulated dc voltage
V_{PRI}	Voltage of T_1 transformer primary
V_{R_1}	Voltage drop across resistor R_1
$V_{R_{BE}}$	Voltage drop across base emitter junction
V_{R_L}	Load voltage
V_{SEC}	Voltage of T_1 transformer secondary

DESIGN PROCEDURE FOR FULL-WAVE CHOKE INPUT FILTER POWER SUPPLY (FIG. 22.)

① Determine load voltage V_{R_L} and dc unregulated voltage V_{DC} (approximately 1.4 V_{R_L} (*GIVEN*).

② Determine load current, I_{R_L} (*GIVEN*).

③ Calculate load power, $P_{LOAD} = V_{R_L} \times I_{R_L}$

④ Choose power transistor Q_1 (with heat sink) to handle 2 × P_{LOAD} power dissipation P_D.

Note: *If* Q_1 *transistor is made of silicon, the base-emitter junction voltage drop* $V_{R_{BE}}$ *is approximately 0.6 vdc. If* Q_1 *transistor is made of germanium, the base-emitter junction voltage drop* $V_{R_{BE}}$ *is approximately 0.2 vdc.*

⑤ Choose a base current for Q_1. $I_B = \dfrac{I_{R_L}}{\text{beta}}$

⑥ Choose a zener diode current $I_{CR_3} = 2 \times I_B$.

⑦ Calculate $I_{DC} = I_{CR_3} + I_{R_L}$.

⑧ Choose a zener diode voltage $V_{CR_3} = V_{R_L} + V_{R_{BE}}$.

⑨ Choose a zener diode to meet ⑥ and ⑧ requirements.

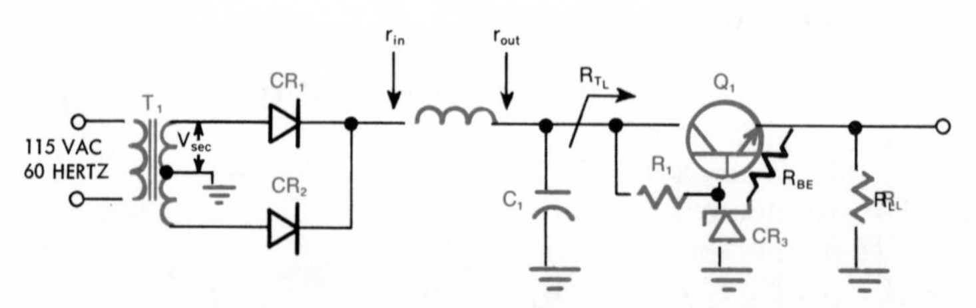

Fig. 22. Choke Input-Filter Regulated, Full-Wave Rectifier Design Schematic

(10) Calculate $V_{R_1} = V_{DC} - V_{CR_3}$.

(11) Calculate $I_{R_1} = I_{CR_3} + I_B$.

(12) Calculate resistor $R_1 = \frac{V_{R_1}}{I_{R_1}}$.

(13) Calculate resistor R_1 power rating $P_{R_1} = V_{B_1} \times I_{R_1}$.

(14) Choose resistor R_1 to meet (12) and (13) requirements.

(15) Calculate total load resistance $R_{T_L} = \frac{V_{DC}}{I_{DC}}$.

(16) Calculate $L_1 = \frac{0.18\ R_{T_L}}{f_r}$ (in henrys).

(17) Calculate $C_1 = \frac{\frac{r_{in}}{r_{out}} + 1}{(2\pi f_r)^2\ L_1}$ (in farads).

(18) Calculate C_1 working voltage $= 2 \times V_{DC}$

(19) Choose capacitor C_1 to meet (17) and (18) requirements.

(20) Choose rectifiers CR_1 and CR_2 (See manufacturers' specifications).
- (a) Reverse voltage rating must be greater than $2 \times V_{DC}$.
- (b) Leakage current (reverse current rating) less than 1 milliamp.
- (c) Forward current rating greater than $2 \times I_{DC}$.

(21) Choose power transformer T_1 (See manufacturers specifications).
- (a) Ensure transformer is 60 or 400 hertz as required.
- (b) Select primary volts (V_{PRI}) as required (110 vac or 220 vac).
- (c) Calculate secondary volts (V_{SEC}) as required. $V_{SEC} = \frac{V_{DC}}{1.3}$.
- (d) Calculate secondary power $P_{SEC} = 1.3 \times V_{DC}$.
- (e) Select T_1 transformer to meet steps (a) through (d).

DESIGN EXAMPLE FOR FULL-WAVE CHOKE INPUT POWER SUPPLY (FIG. 22.)

(1) Given $V_{R_L} = 18$ vdc, $V_{DC} = 1.4 \times 18 = 25.2$ vdc

(2) Given $I_{R_L} = 0.5$ amp

(3) $P_{LOAD} = V_{R_L} \times I_{R_L}$

$P_{LOAD} = 18 \times 0.5$
$P_{LOAD} = 9$ watts

(4) Choose Q_1 to have $P_D = 2 \times P_{LOAD} = 2 \times 9 = 18$ watts.

(5) $I_B = \dfrac{I_{R_L}}{beta}$ *GIVEN:* Q_1 with beta = 50

$$I_B = \frac{0.5}{50}$$

$I_B = 0.01$ amp

(6) $I_{CR_3} = 2 \times I_B = 2 \times 0.01 = 0.02$ amp

(7) $I_{DC} = I_{CR_3} + I_{R_L} = 0.02 + 0.50 = 0.52$ amp

(8) $V_{CR_3} = V_{R_L} + V_{R_{BE}} = 18 + 0.6 = 18.6$ vdc

(9) Choose a zener diode to meet (6) and (8) requirements.

(10) $V_{R_1} = V_{DC} - V_{CR_3} = 25.2 - 18.6 = 6.6$ vdc

(11) $I_{R_1} = I_{CR_3} + I_B = 0.02 + 0.01 = 0.03$ amp

(12) $R_1 = \dfrac{V_{R_1}}{I_{R_1}} = \dfrac{6.6}{0.03} = 220$ ohms

(13) $P_{R_1} = V_{R_1} \times I_{R_1} = 6.6 \times 0.03 = 0.198$ watt. (Use ½ watt standard.)

(14) Choose R_1 to meet (12) and (13) requirements.

(15) $R_{T_L} = \dfrac{V_{DC}}{I_{DC}} = \dfrac{25.2}{0.52} = 48.5$ ohms

(16) $L_1 = \dfrac{0.18\ R_{T_L}}{f_r} = \dfrac{0.18 \times 48.5}{120} = 0.073$ henry

$L_1 = 73$ millihenrys

(17) $$C_1 = \frac{\dfrac{r_{in}}{r_{out}} + 1}{(2\pi f_r)^2\ L_1}$$

$$C_1 = \frac{\left(\dfrac{48}{5} + 1\right)}{(2 \times 3.1416 \times 120)^2 \times 0.073}$$ *GIVEN:* $r_{in} = 48\%$, $r_{out} = 5\%$

$$C_1 = \frac{10.6}{41499.91}$$

$C_1 = 0.0002554$ F

$C_1 = 255.4\ \mu F$

(18) C_1 working volts $= 2 \times 25.2 = 50.4$ vdc

(19) Choose capacitor C_1 to meet (17) and (18) requirements.

(20) Choose rectifiers CR_1 and CR_2 by specifications
(a) Reverse voltage rating $= 2 \times 25.2 \geqq 50.4$ volts
(b) Leakage current rating $\geqq 1$ milliamp
(c) Forward current rating $2 \times 0.52 \geqq 1.04$ amps

(21) Choose transformer T_1 by specification.
(a) 60 hertz
(b) 110 vac
(c) $V_{SEC} = \dfrac{25.2}{1.3} = 19.38$ vac
(d) $P_{SEC} = 1.3 \times 25.2 = 32.76$ watts
(e) Select T_1 to meet steps (a) through (d).

Section 14

Troubles In Solid-State DC Power Supplies

There are many causes for troubles and failures in solid-state dc power supplies but, in general, fewer than in comparable power supplies using tubes. Some of these troubles are simple to analyze because they nearly always occur for the same reasons. For instance, if there is no output when direct current is measured after the regulator, it is apparent that the regulator transistor has burnt out. Some other cases are not so apparent because at least one of several different causes could produce the trouble indication. "Poor regulation" is an example because this could result from three different faults and each would have to be investigated.

Table 1 lists various trouble indications and their possible causes. Generally the defective element should be removed and replaced with a new one, because the attempt to repair these items has doubtful merit. The layman would not have the expertise or the scientific knowledge to repair solid-state devices. He may, with the proper training, rewind a transformer, but the job would take more time than the transformer is worth.

Actual failures may occur with or without any previous trouble indication. The most common failures are described, along with their causes, in paragraphs which follow. Attention to these cannot, of course, remedy the present failure but may prevent similar failures in future applications.

Forward Surge Failures. Forward surges cause failures by raising the temperature of the rectifier in the immediate area of the NP junction. From thermal shock, cracking is caused at the junction. The rectifier then is unable to withstand immediate reverse voltage.

Reverse Surge Failures. Reverse surges cause a rectifier to operate at points beyond its normal maximum reverse voltage limit. Failure is then caused by arcing across the junction and/or thermal shock at the junction.

Mechanical Failures. Failures can happen when the tensile strength of the rectifier's leads is exceeded by pulling, bending, etc.

Mechanical Shocks or Vibration. Each rectifier can withstand only a certain amount of jostling. This is an obvious failure situation.

Thermal Shock. Each rectifier can withstand a certain degree of

TABLE 1. TYPICAL POWER SUPPLY TROUBLE INDICATIONS AND POSSIBLE PROBLEMS.

TROUBLE INDICATION	POSSIBLE PROBLEM
No regulation.	Input voltage dropped below zener avalanche voltage level.
Poor regulation.	1. Power transistor faulty. 2. Zener diode not regulating. 3. Poor quality transformer.
Ripple voltage peaks are uneven in amplitude.	1. Power diodes operating unevenly causing an unbalanced current flow through the power supply. 2. Transformers center top windings are uneven in resistance.
Ripple voltage peaks are not symetrical.	Noise in the zener diode.
Ripple voltage varies in frequency.	AC power supply is fluctuating.
Full wave rectifier power supply indicates half-wave output.	One rectifier circuit inoperative. Isolation of rectifier is necessary.
No output, direct current measured before regulator.	All rectifiers burnt out or circuits open.
No output, direct current measured after regulator.	Regulator transistor burnt out.
No filtering, too much ripple.	Filter faulty.

temperature depending on what it is made of. Glass diodes, for instance, cannot withstand as much temperature as some of the metal assemblies.

Thermal Aging. Constant exposure to high temperature will cause deterioration in diodes. This is similar to hysteresis in metals. Thermal aging can be a mechanical type of situation or can be caused from constant electrical shocks which cause overheating.

Section 15

Power Supply Testing

Testing of power supplies is a simple task. Each power supply is a little different. However, it can be determined by several tests whether a power supply is performing as it was designed to perform. The tests most generally run are output voltage tests, voltage regulation calculations, and voltage ripple tests.

POWER SUPPLY OUTPUT VOLTAGE TESTS (SEE FIG. 23.)

The standard power supply has high and low terminal jacks with which to install the load. This load, of course, has been preestablished by design but can be calculated. Output voltage is checked under no-load and load conditions.

To check the no-load condition, a dc voltmeter is placed in parallel with the power supply. That is between its high and low output terminals. The voltage measurement should be the designed vdc level at which the power supply is rated. As an example, if the power supply is rated at 15 vdc, that should be the voltmeter indication.

Under the load condition a resistor (R_{LOAD}) is used as a load bank. The resistor size is determined by the rated output voltage divided by rated output current. As an example, a 15 vdc, 1 ampere power supply would require a 15 ohm resistor $\left(R = \frac{V_{DC}}{I_{DC}}\right)$. In lab condi-

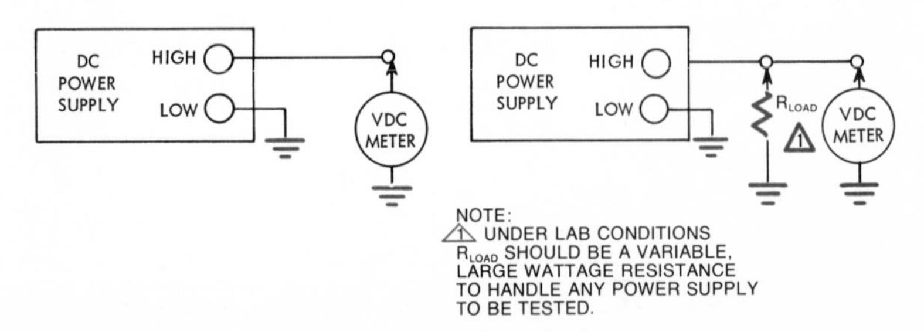

Fig. 23. Measuring the Output of a DC Power Supply

tions a variable resistance is more desireable so that any power supply can be checked. In any event, the power rating of the resistor should be at least $1\frac{1}{2}$ times the volt-amperes of the power supply ($P = E \times I$). The voltmeter is placed in parallel with the load resistor. A good power supply will have the same voltage indication as in load or no-load condition.

If the power supply output voltage with load has dropped to 14 vdc, it would indicate that the load current had dropped from 1 ampere to 0.933 ampere $\left(I = \frac{V_{DC}}{R_{LOAD}}\right)$. This would probably be a result of poor voltage regulation, and the power supply may need to be redesigned.

VOLTAGE REGULATION CALCULATIONS

Let us assume that the power supply no-load voltage was 15 vdc and the load voltage was 14 vdc. The per cent of regulation can be calculated as follows:

EXAMPLE:

$$\text{Percent regulation} = \frac{\text{no-load vdc} - \text{load vdc}}{\text{no-load vdc}} \times 100$$

$$\text{Percent regulation} = \frac{15 - 14}{15} \times 100 = \frac{100}{15}$$

$$\text{Percent regulation} = 6.6\%$$

VOLTAGE RIPPLE TESTS

It is normally sufficient to place an ac voltmeter in parallel with the power supply between the high and low output terminals of the power supply to measure ripple voltage. The indication on the ac voltmeter is the ripple voltage present. The percent of ripple is the ratio of ac to dc voltage at the output times 100.

The use of an oscilloscope will provide a much more accurate look at amplitude of the ripple. It will also supply the frequency and the shape of the ripple. The oscilloscope should be set to ac mode of operation when measuring ripple voltage.

Appendix A
Electrical/Electronic Safety

GENERAL

Safety education today has become an important phase of every training program. Under the 1970 Federal *Occupational Safety and Health Act* (OSHA), the employer is required to furnish a place of employment free of known hazards likely to cause death or injury. The *employer* has the specific duty of complying with safety and health standards as set forth under the 1970 act. At the same time, *employees* also have the duty to comply with these standards.

A full treatment of the subject of safety is far beyond the scope of this book, and there is ample justification for a full course in safety procedures, including first aid treatment, for all electrical and electronic technicians. Instructors should be certified by the state and qualified for any special electrical applications. The intent of this chapter is to make students fully aware of the ever-present, invisible, and generally silent hazards in handling electrical apparatus and to point out some fairly common causes of electrical shocks and fires that can easily be overlooked.

Since the time when Benjamin Franklin flew his famous kite it has become more and more apparent that electricity, even in its milder forms, is *dangerous*. A fact not widely mentioned in history is that shortly after Franklin's experiment a Russian experimenter was killed in his attempt to duplicate the kite trick. We can therefore assume that good safety habits are mandatory for all who use, direct the use of, or come in contact with electricity. Electrical equipment is found in every place that the ordinary person may find himself. For this reason it becomes the responsibility of each of us to be knowledgeable of electrical safety and to become his own and his brother's keeper.

The most basic cause of electrical accidents, as other types of accidents, is *carelessness* and the best prevention is *common sense*. However, knowing exactly what to do in an emergency is only achieved through formal education or experience. Unfortunately, the experience could be fatal, so it is more desirable to derive your knowledge through schooling. When an emergency happens it is often accompanied by panic that can cause the mind or muscles to become paralyzed. The antidote for this panic is education.

RESPONSIBILITY

Electrical/electronic safety is especially important to the technician who is exposed to electrical equipment in the raw state. He is likely to be the one who comes into direct contact with electrical and/or electronic components that may do harm to the body. Since this is the case, he finds himself in the precarious position of being continually alert to hazards that may affect him, his associates, and the people who use the equipment he builds or repairs. The supervisor is responsible for enforcing the rules of safety in the area under his direct supervision. Inspectors ensure that equipment is tested before it is released for use, and final testing should include safety precautions. Finally, the user of the equipment should be qualified in its operation so he may know when it is operating in the proper manner. *No one* who comes into contact with such equipment is exempt from some degree of responsibility.

The average layman may not believe he can get a severe shock from an electrical component disconnected from a power source. The electrical or electronic technician should know better. You know from your study of capacitance, for instance, that capacitors hold electrical charges which they later discharge. A fully charged capacitor disconnected from the power source can deliver a severe and possibly fatal shock, so you take the precaution of "shorting" it before removing it from the equipment. Also, as your brother's keeper, you should warn the user of this hazard.

ELECTRICAL SHOCK

There is a common belief that it takes a great volume of electricity to cause a fatal shock and that high voltage is the thing that provides the jolt to do the job. Imaginative stories about people being literally fried or jolted from their shoes are supplied freely by storytellers. Although there are true tales of this type, this is not generally the case. The bulk of electrical shocks come in small packages. Death from electrical shock happens most often from ordinary 60-hertz, 115 VAC house power. The effect, in general, is an instantaneous, violent type of paralysis. The human body contains a great deal of water and is normally somewhat acid and saline. For these reasons it is a fairly good conductor of electric current, which has no regard for human life or feelings.

Our brains are loaded with tiny nerve ends that provide transmitting service to the muscles. Muscles, in turn, provide motion for the body functions such as the heart and lungs. Now, assume that an electrical impulse that was not called for told your heart to speed up its pumping job or to stop its pumping job. Or suppose that a similar impulse of electricity told the lungs to quit taking in air. These situations do occur and with a comparitively small amount of current flow. Depending on a multitude of complex variables, a small current such as 10 milliamps can be very unpleasant. Currents no larger than 20 milliamps can cause muscle tightening or freezing. Currents of 30 milliamps can cause damage to brain tissue and blood vessels. And damage to brain and blood vessels can, of course, be fatal.

As you know, a decrease in resistance (according to Ohm's Law) causes current to increase. Body resistance decreases when perspiring or otherwise wet. A great number of things can cause variation in resistance. The general health of the person is probably the most important variable. A body in good condition has a much better chance of recovering from electrical shock. The muscles, by being in better tone, can recover to normal from paralyzation.

The path of current flow varies the shock. For instance, if the current involves the brain or heart it is naturally more dangerous than another path. The length of exposure can also be a factor, as well as the size or surface area of the electrical contact. Large voltages can cause spastic action, but recovery can be rapid. Also, small currents can cause muscle paralysis.

In the event of paralysis, artificial respiration or massaging must take place as quickly as possible to prevent loss of body functions and damage to the brain because of lack of oxygen. A condition known as ventricular fibrilation (uncoordinated heart beats, both fast and irregular) can occur with high currents, say 50 to 100 milliamps. This action will continue until something is done externally to restore regular heartbeat.

Death normally occurs when currents reach 250 milliamps. This does not have to occur, however, as rapid first aid can save a victim whose heart has actually stopped beating or whose lungs have stopped pumping momentarily. In many cases rapid action by knowledgeable persons can prevent body damage and save lives.

RAPID RESCUE TECHNIQUES FOR ELECTRICAL EXPOSURE

A special course in first aid will ensure the proper methods. A wrong method can be worse than no action at all. The first rule of thumb for an electrical shock victim is to remove the current path. This can be done by turning the power switch off if it is readily accessible. If not accessible, the person can be detached from the current source by using an insulator of some sort such as wood, rubber, cork, or plastic. Sometimes it is more suitable and sensible to remove the power source from the victim. Whichever is the case, isolation from the current source is by far the most important and first move that can be made.

Isolation procedures can cause problems. The rescuer may find himself in the current path. Touching a person who is paralyzed to a current source provides the current source another path through the rescuer's body. Care must be taken to prevent this from happening. After isolation, artificial respiration and/or other first aid should be applied. In all cases speed is vital. Death occurs in direct proportion to time. It is therefore obvious that a person given artificial respiration in the first three minutes has a much better chance of survival than one who is given artificial respiration after five minutes.

SNEAKY ELECTRICAL CONDUCTORS

In every electrical activity there are sneaky conductors of electricity that cause continuous problems of "shorts" and therefore electrical shock. Anywhere that you have electricity it is not just the wire, the source, or the load that provides these paths for current flow. For instance, cement may seem dry and clean but have moisture in it. In this condition the cement could be a sneaky conductor. Metal floors are, of course, good conductors. A sweating body can cause a multitude of problems. Machines of all types in the general area will serve as conductors. Steel building posts can be conductors as well as metal roofs, steel desks, pots and pans, bicycles, automobiles, refrigerators, washing machines, and just about any other metal object you may name. These sneaky conductors can cause current draw.

Make sure, then, that you are properly grounded or that the equipment you are working with or near is grounded. Use insulated tools to prevent current from flowing where it isn't supposed to. Use floor pads and keep water and oil from floors around which you are performing electrical work. Water does not mix with electricity. Keep debris and scraps picked up to avoid similar situations. Cleanliness and alertness will help avoid or eliminate sneak circuits.

BATTERY HANDLING

Three basic safety problems are associated with the handling of batteries. These are: acid burns, fires, and explosions. Acid burns may be prevented by use of battery-handling clothes and equipment. Clothes for this purpose are mostly made of rubber and include such things as aprons, gloves, boots, and special glasses.

Proper tools are essential to perform the correct procedures safely. Proper flooring will prevent falling and spilling acids.

Fire and explosion may be caused by ignition of gases given off from the charging action. These gases, when mixed with air (oxygen), provide a highly flammable and explosive situation. Gases should not be allowed to accumulate. Ventilators should be installed in battery shops to expel the dangerous gases. Smoking in the area should be prohibited. Signs should be installed warning everyone who might enter the battery shop of the dangers that are within.

HOW TO CONTROL AN ELECTRICAL FIRE

An electrical fire is caused by current flow to some circuit that cannot withstand the current level. Also, electrical fires are caused by sneak circuits which accidently draw current, for instance a "short" to the case in a motor-driven furnace. In any event, since the cause of the problem is current flow, disconnecting the current should be the first step in eliminating the problem. Remove power from the circuit preferably by throwing a switch or by isolating the fire, using insulating material such as wood or plastic. Cut wires with wooden handled hatchets or some similar device. Prevent yourself from becoming part of the circuit. After removing the current, call the fire department, then put out the fire.

Electrical fires are best extinguished with the use of carbon dioxide (CO_2) directed toward the base of the fire. Do not use foam, as it conducts electricity.

GOOD SOLDERING HABITS

Soldering irons or soldering guns all have one thing in common: they are hot. Each is hot enough to melt solder joints. The actual temperature varies with the solder type. The speed at which the soldering iron or gun melts the solder joints is dependent on the wattage of the iron or gun and the size or complexity of the joint. In all events, the soldering device must not only be protected from the handler but also from the other circuits or equipment around it.

The soldering iron or gun should be placed in a heat-sink holder between soldering actions. Heat sinks should also be used to protect electrical/electronic circuit components. Danger of work contamination is always present as dripping or stringing of solder may occur during soldering operations. Fire hazards are always present when working with heat.

Electrical fire hazards may be prevented by ensuring that power is removed from equipment being worked on. Soldering operations should take place only after proper preparation of the work area. Clean and dry work areas, the proper wattage iron, and a well laid out soldering plan help prevent soldering accidents.

Appendix B
Modern Multipurpose Rectifiers and Their Specifications

This appendix is intended to provide the reader a look at some of the solid-state rectifier devices presently available. Each device is illustrated by a photo, installation information in the form of photos or drawings, and specifications in the form of tables.

The examples included are just a few of the many solid-state rectifier devices now on the market, and for this reason the author recommends that this material be utilized for reference only. Before the student or reader designs power supplies, he should also look at other devices offered on the market to find the best device for the job requirement.

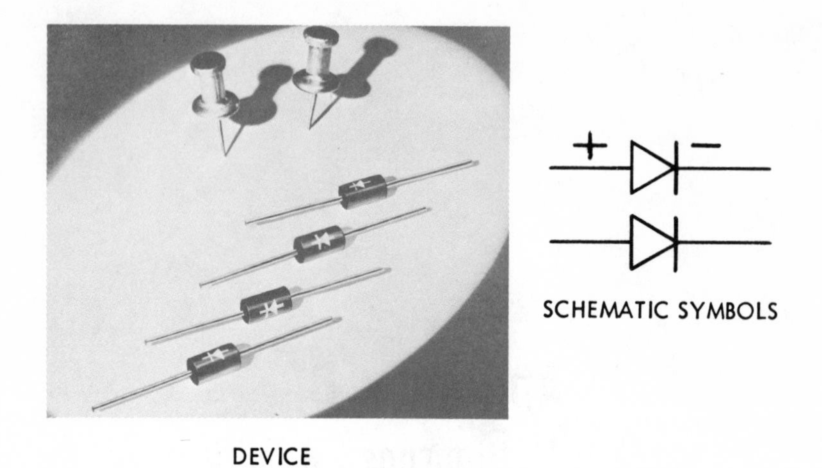

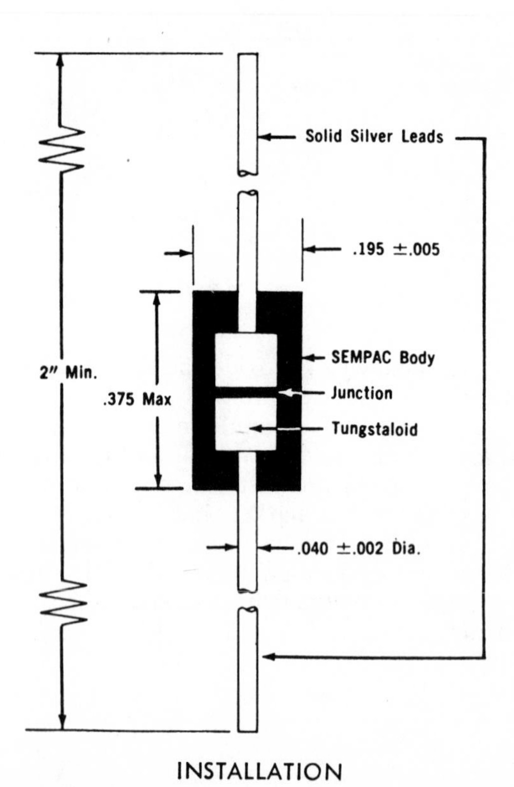

Fig. 1(A). Typical Multipurpose, Axial-Lead Silicon Rectifier (Courtesy of Semtech Corp., Newbury Park, Calif.)

Type No.	Peak Inverse Voltage	RMS Voltage	DC Blocking Voltage	Average Rectified Current		Static Forward Voltage @ 3A		Static Reverse Current		Recurrent Surge	One Cycle	Case
				55°C	100°C	25°C	100°C	25°C	100°C		Surge	
	Volts	Volts	Volts	Amps	Amps	Volts	Volts	UA	UA	Amps	Amps	
IN5197	50	35	50	3	2	1	.95	10	100	30	300	E01
IN5198	100	70	100	3	2	1	.95	10	100	30	300	E01
IN5199	200	140	200	3	2	1	.95	10	100	30	300	E01
IN5200	400	280	400	3	2	1	.95	10	100	30	300	E01
IN5201	600	420	600	3	2	1	.95	10	100	30	300	E01

SPECIFICATIONS

Fig. 1(B). Specifications for Fig. 1(A)

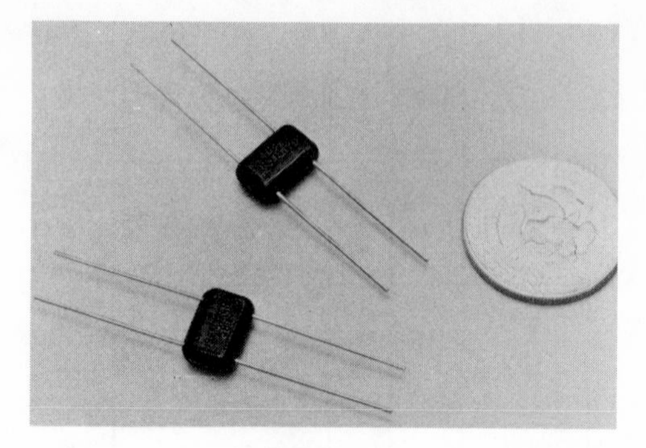

DEVICE

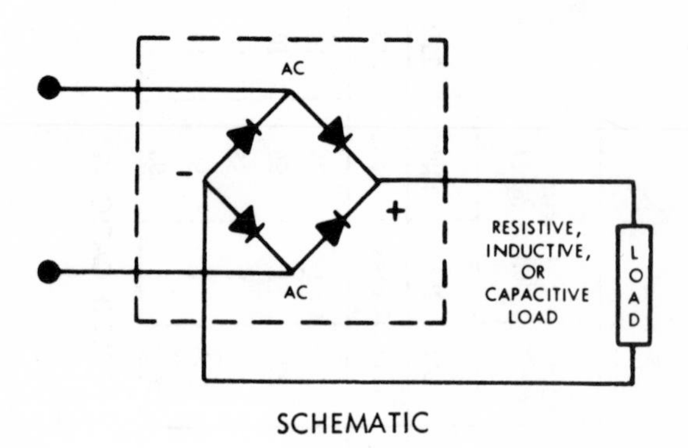

SCHEMATIC

INSTALLATION

Fig. 2(A). Typical Subminiature Bridge Rectifier (Courtesy of Semtech Corp., Newbury Park, Calif.)

Type No.	Peak Inverse Voltage	Average Rectified Current		Max. Reverse Current @ PIV		Max. Fwd. Volts/Leg @ 1 A DC (*100 mA DC)	Recurrent Surge
		55°C	100°C	25°C	100°C	25°C	25°C
	Volts	Amps	Amps	UA	UA	Volts	Amps
SBR05	50	1.5	1.00	2	40	1.2	10
SBR1	100	1.5	1.00	2	40	1.2	10
SBR2	200	1.5	1.00	2	40	1.2	10
SBR4	400	1.5	1.00	2	40	1.2	10
SBR6	600	1.5	1.00	2	40	1.2	10
SBR8	800	1.5	1.00	2	40	1.2	10
SBR10	1000	1.5	1.00	2	40	1.2	10
SBR15	1500	.36	.24	2	40	**3.5	2.5
SBR20	2000	.36	.24	2	40	**3.5	2.5
SBR25	2500	.36	.24	2	40	**3.5	2.5
SBR30	3000	.36	.24	2	40	**3.5	2.5

Solid Silver Leads:
Diameter .031" ±.002" Length 7/8" Min.
Operating and storage temperature
−55°C to +150°C

*Can also be obtained with "Medium" and "Fast" reverse recovery characteristics up to 1000 volts.

SPECIFICATIONS

Fig. 2(B). Specifications for Fig. 2(A)

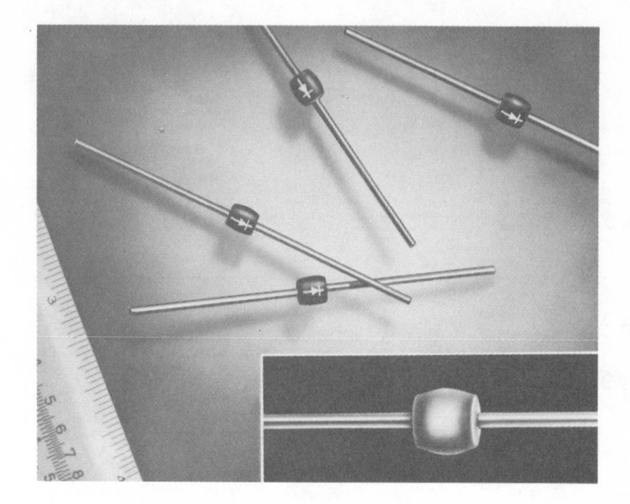

DEVICE

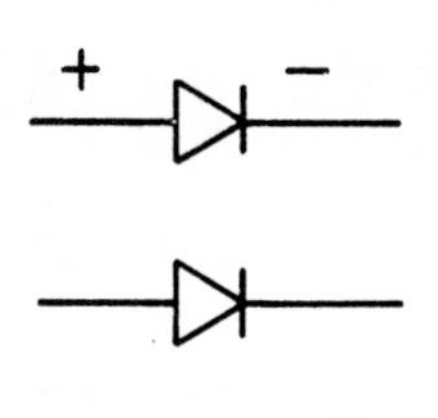

SCHEMATIC SYMBOLS

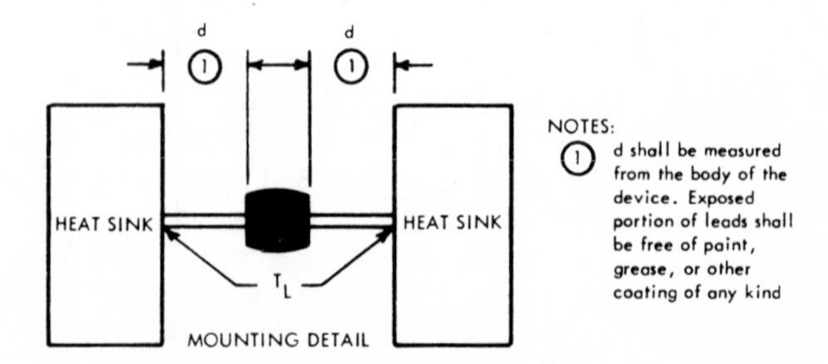

INSTALLATION

Fig. 3(A). Typical Metoxilite Medium Recovery Silicon Rectifiers (Courtesy of Semtech Corp., Newbury Park, Calif.)

Device Type	Reverse Voltage		Forward Current ①	Reverse Current (Max)		Instantaneous Forward Voltage	Repetitive Surge Current	1 Cycle Surge Current tp=8.3 ms	Reverse Recovery Time ②		Typ. Thermal Impedance ③	
			Free Air 55°C	I_R		V_F @ I_F = 3.0 Adc			T_{rr}		Θj–L	
	V_R	V_{RM} (WKg)		UAdc				A (pk)	μs		°C/Watt	
	Vdc	V (pk)	A	25°C	100°C	Vdc Max	A (pk)	I_F (Surge)	Max	Typ	d=0	d=.375
3SM2	200	200	3.0	1.0	20	1.0	25	150	2	1	4	20
3SM4	400	400	3.0	1.0	20	1.0	25	150	2	1	4	20
3SM6	600	600	3.0	1.0	20	1.0	25	150	2	1	4	20
3SM8	800	800	3.0	1.0	20	1.1	25	150	2	1	4	20
3SM0	1000	1000	3.0	1.0	20	1.1	25	150	2	1	4	20

SPECIFICATIONS

Fig. 3(B). Specifications for Fig. 3(A)

DEVICE

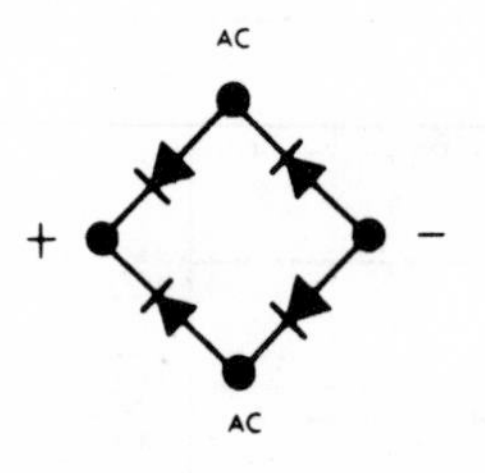

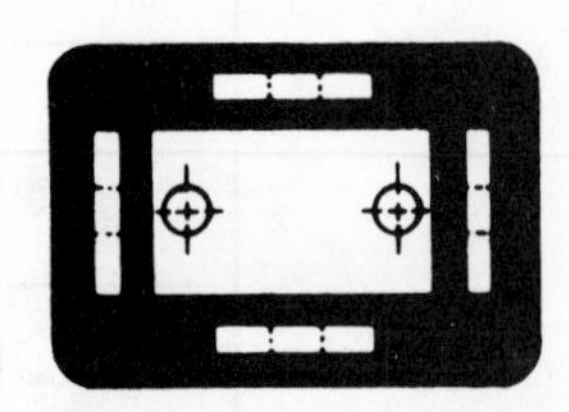

SINGLE PHASE FULL WAVE BRIDGE

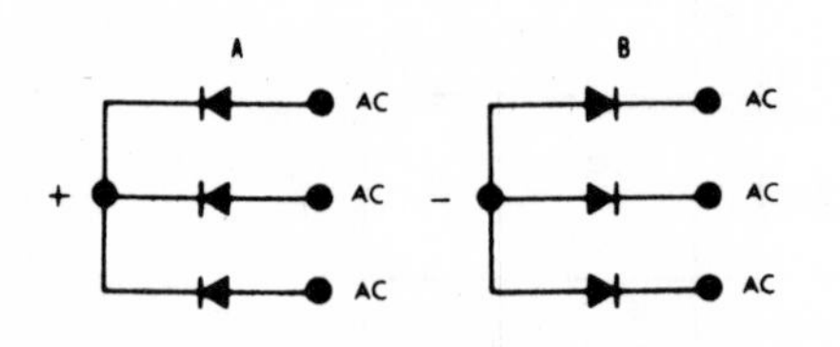

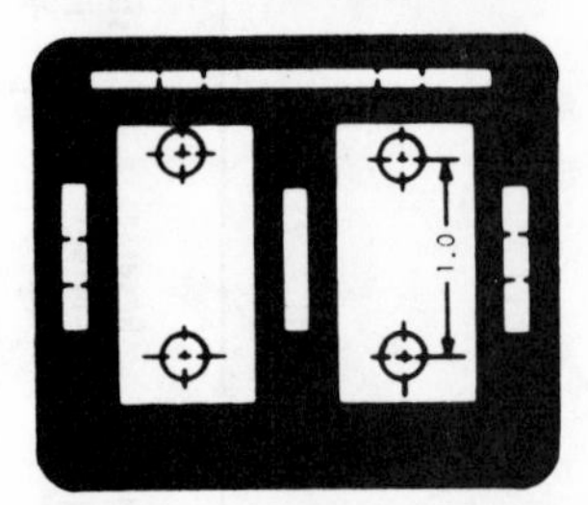

THREE PHASE HALF WAVE BRIDGE

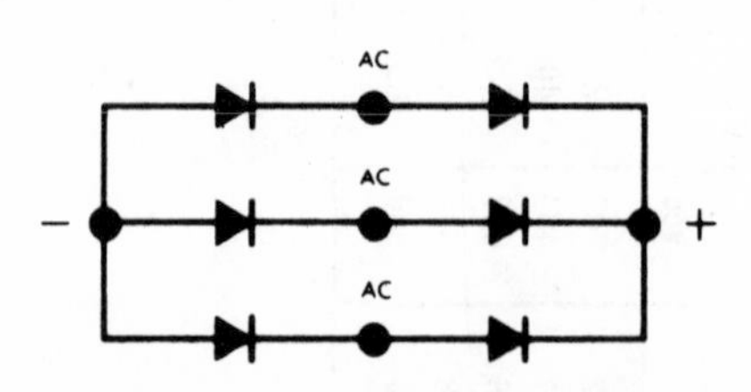

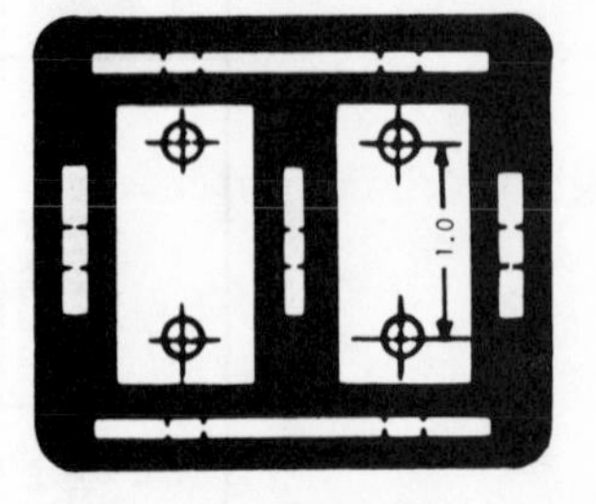

THREE PHASE FULL WAVE BRIDGE

BASE LAYOUT

Fig. 4(A). Typical Single or Three-Phase Bridge Rectifier (Courtesy of Semtech Corp., Newbury Park, Calif.)

Type	No.	Peak Inverse Voltage Per Leg	RMS Voltage Per Leg	Average Rectified Current		Reverse Current Per Leg @ PIV		One Cycle Surge Current	Case Dimensions		
		Volts	Volts	Amps		μA		Amps	Inches		
		25°C	25°C	55°C**	100°C**	25°C	100°C	25°C	H	W	L
1φFWB	SCAS05	50	35	150	80	50	500	900	1.0	1.75	2.5
	SCAS1	100	70	150	80	50	500	900	1.0	1.75	2.5
	SCAS2	200	140	150	80	50	500	900	1.0	1.75	2.5
	SCAS3	300	210	150	80	50	500	900	1.0	1.75	2.5
	SCAS4	400	280	150	80	50	500	900	1.0	1.75	2.5
	SCAS6	600	420	150	80	50	500	900	1.0	1.75	2.5
3φ½WB	SC3HAS05*	50	35	250	130	50	500	900	1.0	2.125	2.875
	SC3HAS1 *	100	70	250	130	50	500	900	1.0	2.125	2.875
	SC3HAS2 *	200	140	250	130	50	500	900	1.0	2.125	2.875
	SC3HAS3 *	300	210	250	130	50	500	900	1.0	2.125	2.875
	SC3HAS4 *	400	280	250	130	50	500	900	1.0	2.125	2.875
	SC3HAS6 *	600	420	250	130	50	500	900	1.0	2.125	2.875
3φFWB	SC3AS05	50	35	190	100	50	500	900	1.0	2.5	2.875
	SC3AS1	100	70	190	100	50	500	900	1.0	2.5	2.875
	SC3AS2	200	140	190	100	50	500	900	1.0	2.5	2.875
	SC3AS3	300	210	190	100	50	500	900	1.0	2.5	2.875
	SC3AS4	400	280	190	100	50	500	900	1.0	2.5	2.875
	SC3AS6	600	420	190	100	50	500	900	1.0	2.5	2.875

*Indicate polarity when ordering
A — Positive Output
B — Negative Output

**Case Temperature

Operating and Storage Temperature −55°C to +150°C

SPECIFICATIONS

Fig. 4(B). Specifications for Fig. 4(A)

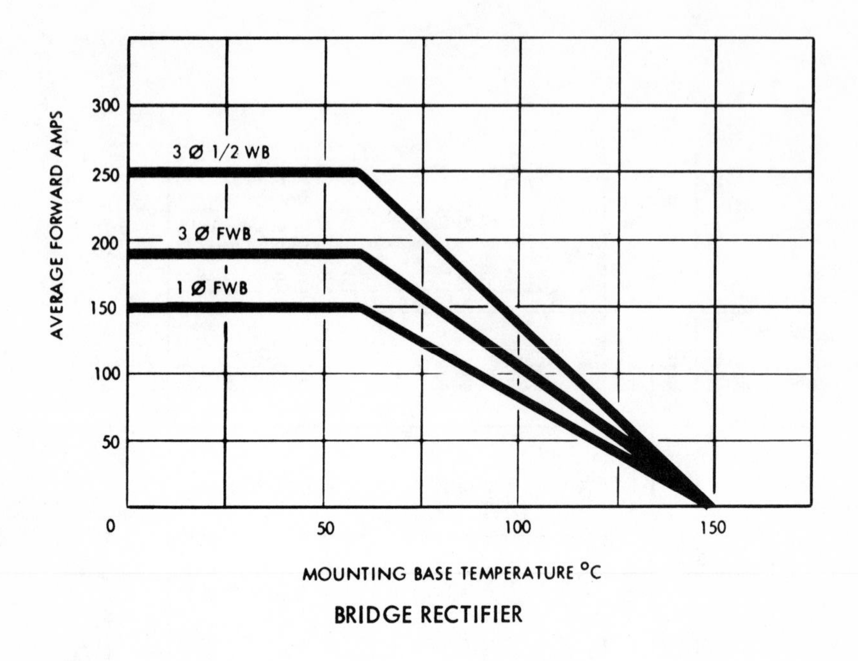

BRIDGE RECTIFIER

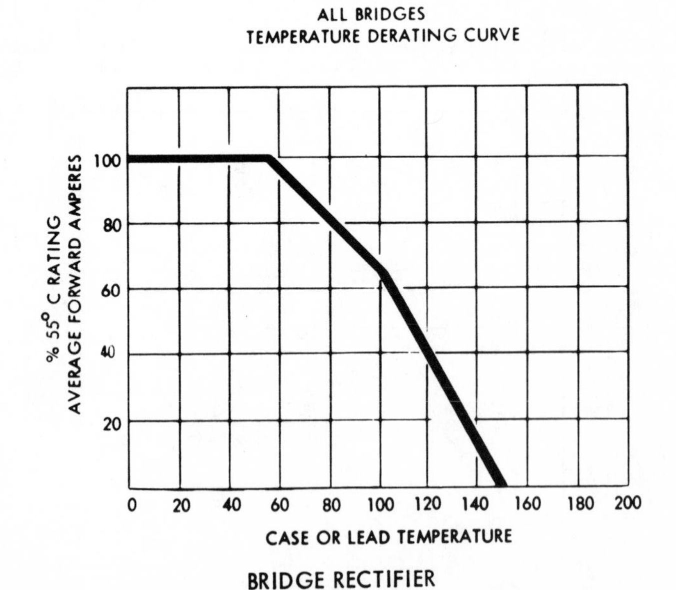

BRIDGE RECTIFIER

Fig. 5(A). Typical Rectifier Specification Curves, Page 1 of 3 Pages (Courtesy of Semtech Corp., Newbury Park, Calif.)

MAXIMUM RATING FOR CAPACITY LOADS

R

C

E IN

$$INSURGE = \frac{E\ IN\ PEAK}{R}$$

INSURGE AMPERES

140
120
100
80
60
40
20
0

25° C

100° C

0.1 0.2 0.4 0.6 1.0 2.0 4.0 6.0 10

RC TIME CONSTANT, MILLISECONDS

SILICON RECTIFIER

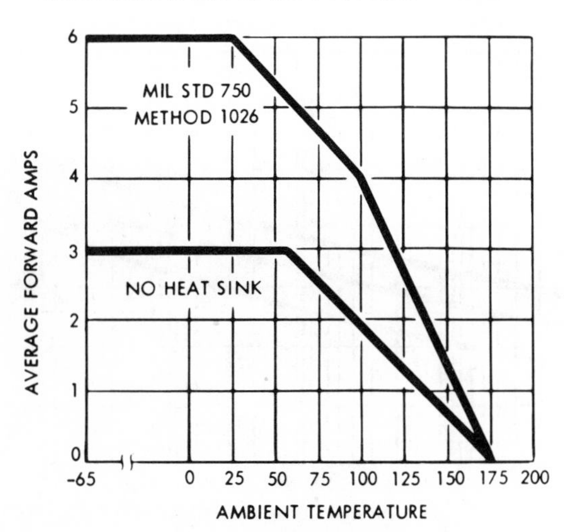

SILICON RECTIFIER

Fig. 5(B). Typical Rectifier Specification Curves, Page 2 of 3 Pages (Courtesy of Semtech Corp., Newbury Park, Calif.)

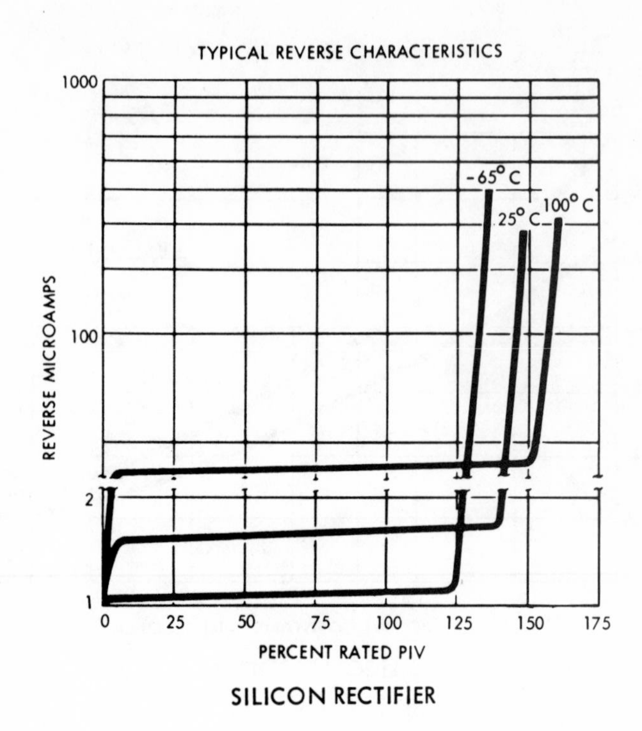

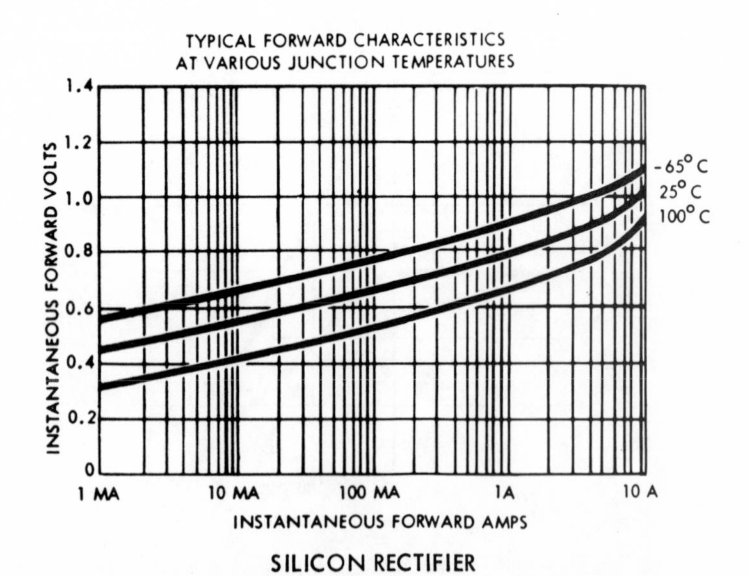

Fig. 5(C). Typical Rectifier Specification Curves, Page 3 of 3 Pages (Courtesy of Semtech Corp., Newbury Park, Calif.)

Appendix C

Typical Modern Power Supplies

This appendix is provided to show the reader a variety of package forms for modern solid-state power supplies. The packages are built to accomplish several goals. Some are quite small to meet the need for light weight and limited space. Some of the larger power supplies have finned heat sinks built on the outside. These are provided for high-power applications where extra cooling capacity from finned heat sinks is needed.

The power supplies illustrated in this appendix are intended to be used for reference only. The author suggests that the reader should bear in mind the thoughts of building neat, lightweight, small, and heat-protected packages with due regard to the power requirement. The same thoughts, naturally, should be in mind if power supplies are purchased.

Fig. 1. Miniature AC to DC Power Module Mounted on Printed Circuit Board (Courtesy of Acopian Corp., Easton, Pa.)

Fig. 2. 3 VDC to 70 VDC Regulated Power Supplies (Courtesy of Acopian Corp., Easton, Pa.)

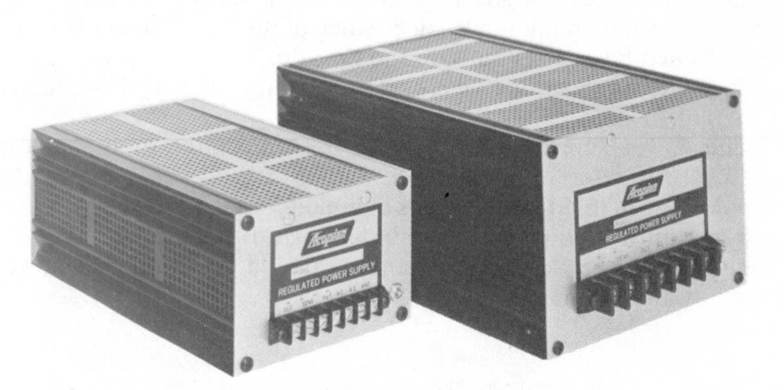

Fig. 3. Adjustable Output, Narrow Profile, Regulated Power Supplies (Courtesy of Acopian Corp., Easton, Pa.)

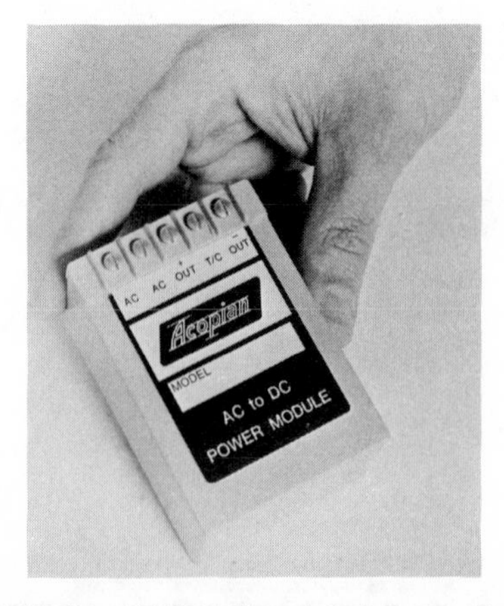

Fig. 4. Miniature 1 VDC to 75 VDC Regulated Power Supply (Courtesy of Acopian Corp., Easton, Pa.)

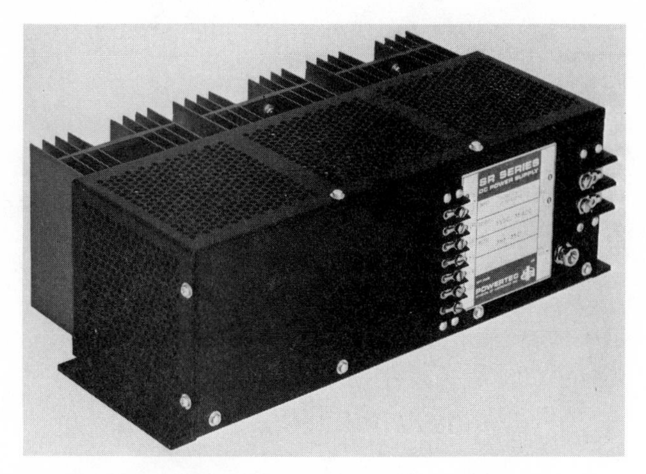

Fig. 5. High-Temperature, Regulated 5 VDC, 35 Amp Power Supply (Courtesy of Powertec, Chatsworth, Calif.)

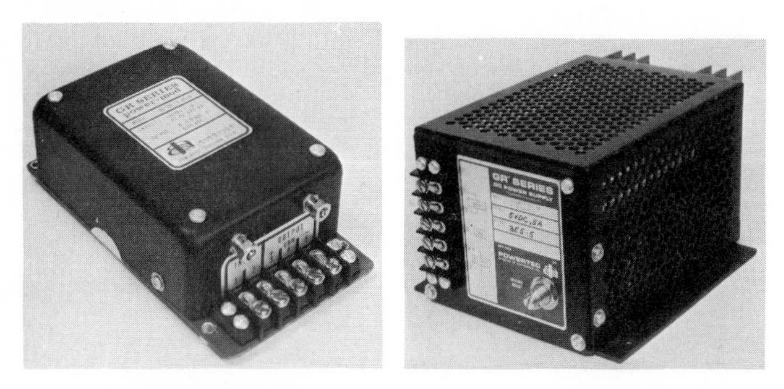

Fig. 6. Plus-or-Minus-Output Power Supplies (Courtesy of Powertec, Chatsworth, Calif.)

Fig. 7. Card-Mounted, Encapsulated Power Supplies (Courtesy of Powertec, Chatsworth, Calif.)